Contested Airwaves

THE HISTORY OF MEDIA AND COMMUNICATION

Series Editors: Jane Rhodes and Lisa Parks

The History of Media and Communication series seeks books that emphasize technology and infrastructure, the politics of social differences and intersectionality, environmental issues, globalization, and labor as they relate to media and communication. Our ambition is to publish monographs and collections that demonstrate innovative historiographic practices and engage with underexplored archives, sources, sites, voices, and experiences. We encourage attention to transnational, postcolonial, and decolonial studies through an historical lens. The series publishes research by scholars in media and communication studies whose work also engages with other fields such as gender and ethnic studies, science and technology studies, environmental studies, and global studies. The goal is to develop original, rigorous, and path breaking books that bring forth the complex pasts of media and communication and that link to contemporary life and near futures.

For a list of books in the series, please see our website at www.press.uillinois.edu.

Contested Airwaves

American Radio at Home and Abroad, 1914–1946

MICHAEL A. KRYSKO

1 2 3 4 5 C P 5 4 3 2 1
♾ This book is printed on acid-free paper.

Library of Congress Cataloging-in-Publication Data
Names: Krysko, Michael A., 1969– author.
Title: Contested airwaves : American radio at home and abroad, 1914-1946 / Michael A Krysko.
Description: Urbana : University of Illinois Press, 2025. | Series: The history of media and communication | Includes bibliographical references and index.
Identifiers: LCCN 2024031453 (print) | LCCN 2024031454 (ebook) | ISBN 9780252046391 (hardcover ; acid-free paper) | ISBN 9780252088476 (paperback) | ISBN 9780252047701 (ebook)
Subjects: LCSH: Radio broadcasting—United States—History—20th century. | Ethnic radio broadcasting—United States—History—20th century. | Foreign radio stations—United States—History—20th century. | Radio broadcasting—Political aspects—United States. | Radio audiences—United States—History—20th century. | Radio programs—United States—Public opinion.
Classification: LCC PN1991.3.U6 K79 2025 (print) | LCC PN1991.3.U6 (ebook) | DDC 384.540973/09041—dc23/eng/20240828
LC record available at https://lccn.loc.gov/2024031453
LC ebook record available at https://lccn.loc.gov/2024031454

To my parents Michael and Irene, my wife Heather,
and my daughter Liliana Zacil

Contents

Acknowledgments ix

Introduction: "The Microphone Is Mightier than the Machine Gun" 1
Visions of Cooperation and Realities of Conflict in Early American Radio

PART I. IMAGINING THE FOREIGN MENACE

1 "Broadcasting in the Language of the Enemies of Civilization" 15
Foreign Language Broadcasting and American Radio, 1920–1940

2 "An Invasion by Radio Is Crossing the Mexican Border" 39
John Brinkley, Border Blasters, and the Geography of American National Identity in the 1930s

PART II. LANGUAGE, EDUCATION, AND IDENTITY ON THE RADIO

3 "To Help the French Speaking People of Louisiana" 65
Language, Education, and Identity in the French Radio Project at Louisiana State University, 1938–1940

4 "An Efficient Way to Spread Shakespeare's Beautiful Language" 90
"Basic English," Language Education, and American International Radio, 1935–1941

PART III. COLONIZED AIRWAVES

5 "A Workable Scheme to Quiet the Panaman Clamor" 117
US Radio Policy in Panama in the Shadow of the World Wars

6 "An Almost Unbelievable Disregard of the Interests of the United States Listeners and Broadcasters" 152
US-Cuban Relations, American Identities, and the 1946 North American Regional Broadcasting Agreement

Conclusion: From "Whistling and Singing 'La Paloma'" to "No Way, José" 171
A Century of Continuity and Change in Communications, Identity, and Borders

Notes 179

Bibliography 225

Index 243

Acknowledgments

The road to complete this book was a long one and people too numerous to count helped me along the way. I owe much to the archivists and librarians who helped facilitate my research at many different institutions. The outstanding people in Interlibrary Loan at Kansas State University, my home institution, consistently tracked down sources essential to this project's completion. Bethany Antos at the Rockefeller Archive Center and Germain Bienvenu at Louisiana State University's Special Collections were especially helpful and responsive in my frantic eleventh-hour efforts to locate, acquire, and confirm permissions to use key photographs that appear in this volume. My gratitude also extends to the *Historical Journal of Film, Radio, and Television* and the *Journal of Contemporary History*, which published earlier versions of chapters 1 and 6 in their August 2007 and October 2018 issues, respectively.

Many friends and colleagues offered constructive feedback and invaluable assistance that helped me turn messy early drafts into the finished chapters that follow. Thanks in particular to Noah Arceneaux, Louise Benjamin, Derek Hoff, Julia Irwin, R. Alton Lee, Yanek Mieczkowski, Roberta Meloni, Jim Schwoch, and Hugh Slotten. I am grateful for the invitation extended by the Leverhulme Research Network's "Connecting the Wireless World Project" that enabled me to participate in its 2018 workshop at the University of Denver. That workshop introduced me to Simon Potter and his scholarship. It also gave me a chance to reconnect with Rebecca Scales, Andrea Stanton, and Derek Vaillant, whose scholarship on radio history has long influenced mine. Mary Elizabeth Kohn, my friend and colleague in the Department of English at Kansas State University, warrants a special shout-out for her insights and input on issues surrounding language and linguistics that run through this book. Finally, the final draft of the manuscript that preceded publication benefited from a last round of input, insights, and suggestions provided by my spring 2024 graduate class on the global history of media and communications. Thank you, Cathy Brewer, Jake England, Dani Ruetti, Kenzie Jansonius, Matthew Self, and Michael Thompson.

Louise Breen and David Graff, both of whom served terms as chairperson of the History Department at Kansas State University while this project worked its way from inception to completion, supported me with travel funding and teaching schedule adjustments that helped facilitate my progress. Deans Peter Dorhout and Amit Chakrabarti were similarly supportive during my own stint as the department's chair. The History Department office managers that I've had the good fortune to work with—first Shelly Reves-Klinkner followed by Melissa Janulis—also helped make this book possible through their efforts in assisting with the travel arrangements that facilitated so much of the essential research and conference attendance. I also owe much to the many conversations I have had with other colleagues, students, friends, and family over the past several years. Not all such exchanges were directly related to the substance of this book, but all helped make it better than it otherwise would have been. Thanks to Christer Ackeroy, Sam Bell, David Defries, Todd Haner, Susannah Krysko, Rich Leslie, Brent Maner, Andrew Orr, Loren Pennington, Joe Scanlan, Rebecca Krysko Scanlan, Steve Smethers, Phil Tiemeyer, Alyn West, Kevin West, and Amy Young.

My parents Michael and Irene Krysko have not only provided a lifetime of support and encouragement, but also allowed me to spend part of my sabbatical during the 2021–2022 academic year at their home on the Five Kezar Ponds in western Maine. It provided an ideal environment to complete the initial rough draft of this book. Soon thereafter I had the opportunity to spend the latter part of that sabbatical polishing and revising that draft on my mother-in-law Judith McCrea's quiet and scenic ranch in Lecompton, Kansas. At the University of Illinois Press, Danny Nasset and Mariah Schaefer did an outstanding job shepherding this project to publication. I am especially grateful for the valuable input and revision suggestions offered by the two anonymous readers, which helped me improve the manuscript in ways I otherwise never could have imagined. Copy editor Renee Cote and the rest of the Press's production team did a fantastic job pushing this book across the finish line. I am thrilled that Lisa Parks and Jane Rhodes deemed the book that emerged from this long journey worthy of inclusion in their History of Communications and Media series.

This point never could have been reached without Heather McCrea, my wife and best friend, who also happens to be my colleague in the Department of History at Kansas State University. She's read many drafts over many years, providing constructive criticisms and insights that greatly strengthened some of the earliest versions of these chapters. Heather's efforts have not just made this book far better than it otherwise would have been but also made me a better scholar. Any shortcomings and flaws in this book, of course, remain my responsibility alone. Whatever strengths it may have are undoubtedly a consequence of the support, help, and feedback I received from so many people along the way, none of whom were more important than Heather.

Contested Airwaves

INTRODUCTION

"The Microphone Is Mightier than the Machine Gun"

Visions of Cooperation and Realities of Conflict in Early American Radio

In the aftermath of World War I, radio inspired widespread popular excitement. This emergent communications technology appeared to many as a much-needed panacea to the recurring division and conflict that too often plagued the world. Broadcasting, which developed as a viable application of radio during that global conflagration, seemed particularly well suited to eliminate the language barriers that evidently hindered cross-cultural communication. Two respected professors of languages at the flagship state universities in Wisconsin and Illinois certainly believed as much. Their 1925 assessment of the quality of language education celebrated a "reawakened interest in a universal language" that "has been quickened by the radio." Such aspirations were, however, not the sole province of academics. That same year General Electric's director of broadcasting, Martin Rice, insisted it took "no strain of imagination" to predict radio would propel the creation of "a universal language" and serve as "the vehicle for complete mutual understanding among the peoples of all civilized nations." Radio working in tandem with motion picture "talkies" had, according to another radio enthusiast, "swept aside all barriers of time and space," which ensured "we shall have a universal language just as surely as we have a universal music."[1]

To others, radio's mere potential to inspire interest in learning foreign languages, rather than producing a universal tongue, could suffice to improve cross-cultural understanding and exchange. Predictably, such aspirations held special appeal to language educators. That prospect particularly excited

C.H. Mercer, a Canadian professor but one whose prominence in the National Federation of Modern Language Teachers ensured his ideas circulated widely among his American colleagues. Thanks to radio, Mercer expected to see an uptick in foreign language interest. With radio supposedly the superior medium in comparison to books and magazines, Mercer postulated that audiences could now "'enjoy' foreign languages from listening to them over the radio." University of Kansas professor of German E.F. Engel concurred. Writing in 1935, he was "quite certain" that radio would "stimulate the study of foreign languages." With proper planning and execution of such educational programming, Engel even imagined radio introducing to "an ever-widening circle of intelligent and appreciative listeners the treasures of the languages and literatures and the contemporary life of people in other lands." Katherine Deveraux Blake, a leader in the National Education Association and peace activist, believed radio's promise was not fully appreciated by an inflexible American educational system. "We study boundaries in geography when planes and radio are setting boundaries at naught and making brothers of the peoples of the ends of the earth," she groused.[2]

The ability of radio communications to traverse boundaries with ease inspired what radio historian Simon Potter calls "wireless internationalism": visions of improving the substance and practice of international relations through the intellectual, cultural, and psychological exchanges across borders. Retired naval commander George Sweet, who had played an outsized role in developing the Naval Radio Services and was still consulted on radio matters well into his retirement, manifested this optimism. In 1923, Sweet expressed his conviction to Secretary of State Charles Evans Hughes that radio would inevitably facilitate "better understanding" between the United States and Mexico, "an appreciation of each other's ideals," and ultimately "the gradual assimilation of each other's language." Not only would a Mexican audience soon develop an appreciation for American radio programs, Sweet believed, but "our own people in turn would soon be whistling and singing 'La Paloma.'" RCA president James Harbord was of a similar mind when he proclaimed in 1927 that communications were the "highways of understanding" that "tie communities and states and nations together more securely than a thousand treaties." Radio, in particular, Harbord continued, promised to pierce the "wall [that] divides two countries from one another," because "understanding comes only with contact." William Castle, an influential diplomat in the Republican presidential administrations of the 1920s and early 1930s, specifically identified radio as "an invaluable channel for the propagation of goodwill, understanding and world peace." If "the pen is mightier than the sword," mused one anonymous writer convinced of radio's global potential, then surely "the microphone is mightier than the machine gun." Nonetheless, this writer did allow that "nationalist politics and prejudice may bar the way for a while." However, because of radio's overwhelming powers of influence, the writer expressed confidence that any such delay "will be a relatively short while."[3]

In fact, nationalism and other embedded prejudices posed not temporary obstacles to these aspirations but enduring ones. When it came to radio, mere contact across boundaries and cultures was not, as RCA's James Harbord insisted, the key to facilitating understanding. Rather, such contact, when infused with nationalism and prejudice, often escalated animosities by exposing listeners to content deemed foreign or otherwise threatening. Across time and geographies, language and accents long served as markers of difference capable of provoking hostile reactions that help illuminate larger societal convictions of social, ethnic, racial, and national hierarchies both within and among nations.[4] It was not that radio, as peace activist and educator Katherine Deveraux Blake presumed, suddenly rendered borders and related signifiers irrelevant. Rather, radio's very ability to cross boundaries at will, irrespective of the jurisdictions they demarcated, often had the opposite effect: underscoring the permeability of borders that were expected to separate different cultural and national communities. The medium's seeming disregard for internationally recognized borders often provoked fears and anger about their vulnerability to invisible radio threats. Those most vexed by unwelcome radio trespasses frequently drew from notions of national and other forms of personal identity to challenge a perceived radio threat and those responsible for it. The point here is not just that the utopian-like expectations and predictions for radio in the aftermath of World War I fell far short of realization. It is instead to illustrate that the very ease of cross-border contacts often had the contrary outcome; the way people used radio facilitated misunderstanding and destabilization, intensified deeply rooted animosities, and solidified existing divisions often rooted in identity, nationalism, and prejudice.

This divide between expectations and outcomes is hardly unique to radio. In 1846, for example, one newspaper editorial confidently proclaimed that the telegraph would ensure Americans "shall become more and more one people, thinking more alike, acting more alike." The telegraph would, this piece concluded, allow Americans to share "one impulse" and "man will immediately respond to man."[5] In fact, in the fifteen years that followed that prediction, the telegraph circulated news that antagonized the North-South sectional divisions, helping drive the United States into its Civil War. Men certainly did respond to each other, albeit with a shared impulse for violence and aggression not anticipated by the original quote. International telegraphy in the latter part of the nineteenth century also fell short of providing the expected salutary benefit to diplomacy. For many diplomats accustomed to exchanging and considering messages over a span of days, weeks, or even months, it became a struggle to adapt to the accelerated speed of diplomatic communications and accompanying demands for rapid response. For example, a dispute that followed the 1891 death of an American sailor on Chilean soil nearly led to war, due in no small part to both countries' hot-headed leaders' inclinations to use the telegraph to convey ill-considered confrontational missives.[6]

These patterns continued into and throughout the twentieth century. Stephen Kern makes a compelling case that during the summer of 1914 pressures for immediate diplomatic responses via the telegraph intensified the disagreements and divisions propelling Europe into World War I. During the Cold War, satellites did not, as President Lyndon Johnson predicted, inspire a "partnership" among the United States, the Soviet Union, and "all nations under the sun and stars." Space communications policies under Johnson and his predecessor John F. Kennedy instead became another of the many focal points of Cold War international rivalries. The internet was (and at times still is) the focus of the same timeworn predictions that its development promised to facilitate peace, unity, and mutual understanding. Such predictions stand in contrast to the internet's frequent use as an abundant provider of misinformation, propaganda, and outright lies by both foreign and domestic actors seeking to antagonize and widen existing partisan and cultural divisions.[7]

With a focus on radio in the era of the world wars, the chapters that follow explore this divide between the technological enthusiasm for communications that greeted radio in its earliest years and outcomes that frequently belied such utopian predictions. In so doing, it builds on the so-called transnational turn that has infused the historical writing of US history more generally and radio studies more specifically. Casting his net broadly, historian Thomas Bender called for an understanding of American history recognizing that national history reflects a convergence of forces from both within and beyond the nation. "The nation is not freestanding and self-contained; like other forms of human solidarity, it is connected with and ultimately shaped by what is beyond it," Bender advises. This transnational perspective is increasingly well represented in radio history, underscored by radio scholar Derek Vaillant's counsel to consider "international connectivity as central, rather than peripheral, to the rise of modern broadcasting." Doing so, he argues, "can shed light on formation of radio nations, that is, nationally bound broadcast cultures," which invariably reflect the dynamic influences of regional, national, and international cultural and technological networks. Vaillant, whose study explores the many Franco-American radio exchanges from the 1930s through the 1960s, documents how the increased contact during the late 1930s illuminated the differences in the two countries' national broadcasting approaches, styles, and audience preferences.[8]

The ease with which radio communications crossed international borders has made it a subject well suited to this transnational pivot in considering unrealized aspirations for wireless internationalism. Despite giving lip service to the rhetoric of wireless internationalism, the BBC established itself as a megaphone for Britain's decidedly nationalist foreign policy objectives on the eve of World War II. In the same period, pan-European efforts to develop intracontinental broadcasting based on technical considerations and expert recommendations foundered in the face of resistance fueled by national politics and international rivalries. French colonial broadcasting practices in the 1930s did not, as hoped, create a united empire; those practices instead did more to underscore perceived distinctions between the presumably civilized metropole and the exotic, dangerous, and uncivilized French colonial periphery. Also in the 1930s, the Palestinian Broadcasting Service helped foster and magnify the separatist national identities of Arab Palestinians and Zionists, thus adding fuel to their enduring conflict. In the run-up to the Pacific War, American radio initiatives in East Asia that claimed internationalist aspirations instead provoked bitter conflicts in US relations with both China and Japan. Well before that war pit the United States against Japan, Japanese bureaucratic rivalries and parochial infighting sabotaged Japan's efforts to develop an expansive international telecommunications network that tied together its far-flung but ill-fated empire. Japan's defeat in that Pacific War further exposed the significant shortcomings of the sprawling but flawed network.[9]

This book seeks to build on such scholarship and further explore the dynamics of radio's divisiveness in multicultural and politically volatile contexts. To do so, it prioritizes a focus on the ways in which radio's aural attributes engaged listeners; through this engagement, radio encourages its users to draw on their existing knowledge, assumptions, and prejudices to reach often suspect conclusions about the allegedly "foreign" or otherwise unwelcome content they heard. Radio's aurality provokes listeners to imagine a visual context that surrounds the disembodied voices and sounds coming from the speaker. Such imaginings, according to cognitive psychologists, are among the most powerful, emotional, and enduring a mind can create. Unlike visual media, in which images are integral to the content, aural media compels the mind to apply extra effort to imagine the seemingly relevant visual context when none has been provided. The resulting visualizations of aural content are shaped by a listener's existing body of knowledge and assumptions. The relevant body of knowledge and assumptions that inform this visualization process is inextricably intertwined with the historical specificity of the listening experience; how content is visualized and understood reflects one's position within the larger social, cultural, and political fabric of the surrounding context. From that vantage point, these cognitive processes will invariably draw from that historical moment's relevant discourses and ideas surrounding, for example, race, gender, class, or any other lens that seems relevant to making sense of the content in question.[10]

Collectively, this visualization process and the reactions it elicits help explain why listeners will often make assumptions about the looks, motivations, and intelligence of speakers based on accents or other sounds rather than any substantive content conveyed. Depending on the listener, the same content might draw different reactions from those listening from different societal vantage points. For example, in the 1930s many educated middle-class white listeners found the highly popular but racist and stereotype-laden program *Amos 'n' Andy* humorous in its depiction of African Americans, while many Black listeners from the same period predictably found the program derogatory and offensive. Still, other Black listeners professed enjoyment for *Amos 'n' Andy* because it showcased Black protagonists with their own agency. The importance of historical specificity of the listening experience is further underscored when considering that many in a demographic of white middle-class listeners hearing the program in our current era would likely be flabbergasted by the blatant racism infused into the program that their 1930s predecessors found wholly acceptable and unremarkable. In this way, varying reactions to content offer a window into listeners' situationally specific sense of self and identity in their historical moment.[11]

Finally, such visualization processes occurred against the backdrop of emergent views that the airwaves were tangible parts of national territory. The so-called ether across which radio transmissions traveled was, in this conceptualization, part of a nation's natural resource endowment to which that nation was singularly entitled. It was invisible to the eye and unable to be touched. And yet, this ether was increasingly conceived as being as much a part of a country's tangible natural resource entitlement as were the forests, waters, minerals, fossil fuels, and other valuables that also lay within the nation's prescribed territorial boundaries. "Keeping the Stars and Stripes in the ether," as Naval Radio Division head Stanford C. Hooper put it in 1922, was essential to keeping foreign radio trespasses at bay. In short, this view of the airwaves as an integral part of tangible territory gained traction at a time when radio's aural attributes could encourage listeners to visualize disagreeable content as somehow foreign and inappropriate. In the many instances explored in the pages that follow, it was not just that listeners deemed such content disagreeable, but that this content reflected the unwanted, unwelcome, and threatening territorial trespasses of an outside interloper.[12]

These realities worked at cross-purposes with idealistic visions of radio's potential to propel mutually beneficial understanding and exchanges across borders, cultures, and societies. Such convictions were a manifestation of a simplistic but deeply rooted technological determinism, which often warped popular expectations and understandings of many new technologies' presumed power and potential. Claims that radio could singularly annihilate borders, propel the creation of a universal language, or bind nations together more effectively than peace treaties reflected such deterministic thinking. Those deterministic assumptions ignored the historical specificity of the listening experience and ignored the plurality of listening positions into which individuals of a mass audience fit; in so doing, this technological determinism ignored the critical factors that could encourage different listeners to have contrary reactions to the same programming content. Instead, as social constructivist approaches to the history of technology underscore, the trajectory of technological development and use is shaped by human-driven calculations and choices, all of which are made within the parameters of the surrounding political, economic, social, cultural, and legal contexts that buttress convictions of what is desirable and possible, as well as what is undesirable and worth avoiding. Adverse reactions to perceived foreign or other inappropriate outsider radio intrusions thus offer compelling examples of political and cultural factors shaping listener engagement with radio.[13]

The chapters that follow explore different permutations of that fundamental premise. The explorations are girded by three related bodies of radio history scholarship. By the first decade of the twenty-first century, studies written from the vantage points of both communications studies and history of technology explore the array of commercial, political, technological, and cultural interventions that shaped the regulatory practices, programming choices, and listening habits of what became mainstream American broadcasting.[14] Collectively, these studies show how such interventions transformed the poorly regulated US airwaves flush with a diversity of programming in the early 1920s into the commercialized, advertising-supported, and network-dominated broadcasting system that by the start of the next decade became widely accepted as a distinctly "American system" of radio. Complementing this scholarship is the rich body of work that emerged since the early part of this century, which looks beyond mainstream radio and identifies the significant gaps in coverage, exceptions, and resistance to the so-called American system.[15] This subsequent generation of scholarship underscores that American radio was not wholly subsumed by commercialization, while support for the commercialized American system was far from universal. Finally, accounts reflecting the recent "transnational turn" in radio history writing further illustrate the significance of an underappreciated array of global influences and international connections that infused purportedly distinct American broadcasting practices.[16]

From this historiographical foundation, substantial attention is placed on the listeners and policymakers who accepted and embraced this American system while objecting to practices, policies, and actors who appeared to challenge it. The first two chapters, comprising the book's first part, focus on the perceived threats that foreign interlopers posed to American radio. Chapter 1 considers listener reactions to US-based broadcasts in languages other than English during the 1930s and early 1940s. With the purported American system of broadcasting solidifying its position as the expected norm at the start of the 1930s, protesting listeners alleged these foreign language broadcasts were antithetical to what should be allowed on the American airwaves. With wars being waged in Europe and Asia by the end of the decade, concerns about foreign disloyalty and "fifth-column" activity began to infuse the objections. These reactions manifested expressions of national identity, illuminated the place of spoken language within that identity, and claimed the ether as a territorial space deserving protection from outsider intrusions.

The second chapter redirects the focus to perceived foreign threats coming from beyond US borders, specifically Mexico. It too underscores the strong convictions held by many listeners regarding the proper and appropriate use of the airwaves as an identifiably American space. But in this instance, the purveyor of the threat was the English-speaking American quack doctor and broadcasting provocateur John Brinkley. Ousted from his perch on US radio amidst its commercialization and stricter licensing requirements, Brinkley was seen by the Mexican government as a diplomatic weapon that could be used against its long-domineering northern neighbor. Mexico encouraged Brinkley to disrupt American radio for the purpose of pressuring the United States to relinquish its stranglehold over the hemisphere's radio frequencies. US diplomatic stubbornness and incompetence loomed large in enabling Brinkley to remain on the air from Mexico between 1931 and 1941. During that period, outraged listeners, armed with the knowledge that these broadcasts came from beyond US borders, demanded their government protect the American airwaves from attack. These listeners' propensity to imagine otherwise invisible transmissions originating in Mexico before violating US territory fueled perceptions of these largely English language broadcasts as real and present threats to the nation.

The book's second part, comprised of chapters 3 and 4, considers two educational initiatives that received funding from the philanthropic Rockefeller Foundation and operated outside the mainstream of US commercial broadcasting. Chapter 3 heads to Louisiana State University, where a radio-based French language project developed in the late 1930s sought to educate Louisiana's large poverty-stricken and poorly educated Cajun population. It was a community that traced its origins to an initial migration of French Canadians into the Louisiana territory during the late eighteenth century followed by subsequent global migrations of many other non-Anglos to the region by the twentieth century. These successive migrations collectively shaped the development of a distinct Cajun culture and language in the context of rising Anglo-American dominance in the larger United States. Harley Smith, the linguist who developed the project, advocated using the Cajun dialect of French in the scripts to maximize possibilities of engaging and educating the intended audience. His efforts were undermined and the project ultimately collapsed when French language teacher Louise Olivier, the other principal leader, forced Cajun out of the scripts in favor of standard French. Olivier, a Cajun herself, was invested in an elitist myth-laden view of Louisiana's supposedly purer but rapidly vanishing French-inherited traditions to which the inferior Cajun dialect posed an existential threat. Olivier's intervention resulted in a radio project poorly suited to educating its intended Cajun audience.

A contemporaneous effort to use shortwave broadcasting to teach a simplified form of English to audiences across Latin America and the Caribbean suffered from comparable shortcomings. Just as a misguided sense of "French exceptionalism" infused the LSU project, the so-called Basic English educational initiative, the subject of chapter 4, was suffused with misplaced notions of American and Anglo-Saxon exceptionalism. The broader history of US foreign relations is rife with examples of exceptionalist convictions justifying the pursuit of American territorial, economic, and cultural expansionism as serving a greater global good. By the early twentieth century, an Anglo-Saxon-specific derivative of that exceptionalism was often used to explain appearances of an increasingly close US-British relationship amidst the outsized global power both countries collectively exerted. The Basic English project, which was itself a collaboration between British and American principals at a Boston-based nonprofit shortwave broadcaster, is an example of this exceptionalism at work and the extent to which its cultural baggage can work at cross-purposes with achieving the goals at hand. The project was undoubtedly sincere and well meaning. However, its architects gave insufficient consideration to how the intertwining of identity, spoken language, and ideas of the ether as protected national space affected how target audiences across Latin America and the Caribbean might engage with broadcasts that originated from well beyond their national boundaries. That blind spot combined with their inattentiveness to the region's larger linguistic diversity circumscribed the potential of the Basic English radio project to engage a vast mass audience of non–English speakers across the Western Hemisphere.

Part III, encompassing the book's last two chapters, is situated at the nexus of radio, US foreign policy, and American imperialism and colonialism. The blindness to larger Latin American perspectives that undermined the Basic English project is, in fact, an overarching theme in the larger history of US foreign relations that infuses the foreign policy–focused third part of the book. The subject of chapter 5 is Panama, across which the United States built a transoceanic canal at the start of the twentieth century. As part of its deal to build and operate the Panama Canal, the United States administered a colony, the so-called Canal Zone, that cut across the country's midsection. The treaty that underpinned US authority over the Canal and surrounding zone was so broadly written—deliberately so—that it effectively subordinated the rest of Panama into an informal US colony. That expansive authority provided the basis of the US demand that all radio in Panama, even within Panamanian jurisdictions, be placed solely under American control. Such control was necessary, US officials argued, to ensure the effective functioning of the communications systems that facilitated the operation and protection of the Canal. While many Americans recoiled at perceived foreign radio intrusions, Panamanians strenuously objected to the actual domination of their airwaves by the United States. From the 1910s through the 1940s, intransigent and short-sighted US policies utterly failed to prevent Panama from ultimately asserting control over radio outside the Canal Zone. Instead, ill-conceived approaches to radio, spearheaded by the US Navy, mostly served to fuel Panama's ever-increasing anti-American nationalism. In conjunction with other heavy-handed US actions and policies, military and diplomatic overreach only served to illustrate the limits of US power, and ultimately weakened the overall American position in Panama over the course of the twentieth century.

A radio rivalry with Cuba is a focus of the sixth and final chapter. It was a rivalry borne of the Americans' colonial legacy in that country. Cuba was occupied by the United States at the start of the twentieth century. It was long governed by a sovereignty-limiting constitution the United States imposed on it in 1901 as a condition of ending that occupation. In the years that followed, Cuba was effectively rendered an informal colony due to the preponderance of economic and political power the United States exerted over the small nation that lay just ninety miles off Florida's coast. As a result of this smothering American presence, anti-American nationalism had emerged as a powerful force in Cuban politics by the 1940s. The inundation of the airwaves with American programming, leaving little room for Cuba to develop broadcasting for its people, further incited these anti-American frustrations. In 1946, Cuba set its sights on the 1941 North American Regional Broadcasting Agreement (NARBA). That treaty resolved the earlier US-Mexican border radio dispute that is the focus of chapter 2. Taking a page from a playbook that served both Mexico and Panama very well, Cuba gave notice that if the 1941 NARBA agreement was not revised to assign Cuba more frequencies, it would wreak havoc on American broadcasting through static interference. With lessons seemingly learned from Mexico and Panama, the United States quickly agreed to a revised NARBA that satisfied

Cuba's concerns. That decision, however, provoked farmers from across California and Arizona to rail against the US government's acquiescence. Protesters claimed that the sacrifice of one particularly important frequency, that of the powerful Los Angeles-based station KFI, threatened their access to vital news and information that dictated how they tended to their crops when faced with looming inclement weather. These dissenters bitterly protested the likelihood of interference, as well as the prospect of hearing Cuban "rumba" and other Spanish language broadcasts on that frequency. Their objections manifested a distinct Far Western regional identity that was a part of their national identity, and in so doing this conflict helps illuminate the importance of region in the larger history of identity formation in the United States. At the same time, the promise of radio-inspired internationalism remained unrealized with listeners once again resisting and objecting to perceived foreign intrusions on supposedly national airwaves.

This book, at its core, is about exaggerated and unrealistic convictions about radio's inherent power to affect change, for good or bad. Aspirations for radio to unite the world in peace and mutual understanding were no more realistic than hyperbolic warnings of dire threats to the nation manifested by purportedly malfeasant radio users. Each chapter investigates a story in which control over content and access to the airwaves became a focal point for the pertinent listeners, broadcasters, and policymakers. Each story places a focus on how the principal actors conceived of content as appropriate or inappropriate for radio, and then worked to allow or deny access to the airwaves for such content. Those efforts typically reflected the underlying intersection between identity, language, politics, and radio's presumed power to influence and affect change. How people made sense of and chose to act on those intersecting ideas predictably varied across different individuals and demographics. Such variations are the reason radio, at least in the topics explored here, became a flashpoint prone to sow division and animosity, both domestically and internationally. The mutual understanding and cultural unity so many earnestly predicted that broadcasting promised to facilitate remained elusive.

Some caveats about sources and their limitations are required. The chapters exploring the radio conflicts with Mexico, Panama, and Cuba, as well as the hemispheric-wide "Basic English" broadcasting initiative, necessitated drawing conclusions about motivations, reactions, and decisions of some Latin American actors from a limited source base. For this task, the book relies on a relatively small number of published Spanish language sources and newspaper articles, translations of documents preserved in the US governmental archives, and a rich body of secondary works in Latin American history and US-Latin American relations, while striving to avoid claims and conclusions that overstep what this source base supports. While this source base has evident limitations in providing a complete picture of the Latin American perspective, I believe it nonetheless provided sufficient evidence and information about the worldviews, assumptions, and convictions that were central to shaping the ideas about, engagement with, and policies for radio.

The analyses are also mindful of an evident limitation when it comes to drawing conclusions about listener engagement with radio. Radio historian Jason Loviglio described the challenge of doing radio history as finding those "moments when the audience becomes visible."[17] The seemingly voluminous listener correspondence and reactions culled from newspapers, radio magazines, US State Department documents, Federal Communications Commission records, the Rockefeller Foundation, and other scattered archival sources, in fact, reflect only a miniscule portion of the total listening audience at any given time. The typical listener, whether inclined to offer praise or harboring a grievance, did not take the time to commit their sentiments to writing, much less pay for the postage to relay them via the mail to a recipient. Reports from government, business, educational, and philanthropic officials who, in addition to being listeners themselves, made radio policies and pursued radio initiatives, help broaden that source base to some extent. Nonetheless, it still falls far short of providing a representative sampling of the radio users under consideration. The challenge, then, is to be mindful not to draw overly sweeping conclusions about expectations, patterns, and reactions that exceed the available evidence.

For this reason, the fundamental argument makes no pretense toward claiming unifying and all-encompassing convictions about radio rooted in a singular national identity or consensus about appropriate content, whether in the United States or abroad. Absent claims of some larger all-encompassing listener or radio experience, the value in the analyses running through the following chapters squarely rests on the diversity of opinions and reactions they collectively illuminate. As with all mass media, radio was not and never could be a uniformly homogenizing communications medium. The people and groups under consideration here engaged radio by filtering the content and ideas they encountered through their own individual worldviews, assumptions, and convictions. Such filtering, in turn, underpinned outlooks, decisions, and actions capable of propelling misunderstanding and conflict when confronted by personally disagreeable content. In this regard, this book's source base is indeed useful in the effort to draw meaningful conclusions about how a variety of radio users from different backgrounds and experiences understood the medium within the larger context of radio's development in the United States and abroad during the first half of the twentieth century.

PART I

Imagining the Foreign Menace

CHAPTER 1

"Broadcasting in the Language of the Enemies of Civilization"

Foreign Language Broadcasting and American Radio, 1920–1940

Kingston, Pennsylvania, resident Earl Andrews wanted to expunge foreign language broadcasting from the US airwaves. He made this view clear in a 1931 complaint he sent to the Federal Radio Commission (FRC). "As I understand," an incredulous Andrews asserted, "every citizen of the United States must and should understand the English language." And those who were not American citizens, according to Andrews, "should not be considered." Broadcasting, without equivocation, should only be in English. "I see no necessity of using language other than the English language in these United States," the outraged complainant concluded. "I feel that the language of such men as George Washington, Thomas Jefferson, Abraham Lincoln, Theodore Roosevelt, and Woodrow Wilson is the language for me or any other citizen of the United States of America."[1]

Andrews's diatribe captures the central role English played in defining popular conceptions of American national identity. The English language long claimed a prominent role in historical conceptualizations of American identity. This connection between language and identity intersected with historically fluid, oft-debated notions about the potential of and processes for immigrant assimilation into American society. Andrews articulated his complaints in the context of broadcasting's commercialization that by the 1930s had already drastically reduced the number of foreign language programs on American radio. This commercialization popularized the notion of a distinctly American model of radio broadcasting dominated by new networks like the National Broadcasting Company (NBC) and the Columbia Broadcasting System (CBS). By the

1930s, network programs enjoyed a national reach and the means to broadcast the same program to an expansive nationwide audience. In this regard, commercialized broadcasting was well situated to engage listeners' sense of their American national identity. This commercialized transformation of American radio rendered foreign language broadcasting far more vulnerable to criticism by the 1930s, even though its presence on the airwaves had markedly diminished.

Positioned as an outlier on the radio dial in the wake of that commercialization, foreign language radio programming in the United States provided listeners like Andrews an opportunity to articulate their sense of national identity and perceived entitlements as American citizens. Provoked by foreign language broadcasting, many self-described patriotic listeners insisted that American radio be purged of any allegedly un-American presence. Such programs managed to capture the attention of listeners who did not speak the respective languages or understand the broadcasts. Even the famous entertainer Groucho Marx felt compelled to chime in with his grievances. By articulating their strong opposition to foreign language programs, many Americans expressed their sense of belonging to an American community through nationalistic and, at times, xenophobic protests designed to minimize any non-English (and therefore un-American) presence on US airwaves. Embracing newly emergent conceptualizations of true "American" radio, aggrieved listeners railed against ominous unknown foreign language communications that traveled invisibly through the air and purportedly threatened the very fabric of American life.

The Marginalization of Foreign Language Radio in Interwar Era America

Earl Andrews's 1931 grievance belied that a decade earlier American radio listeners often embraced foreign language programming. In 1925, commercial broadcasters owned only 4 percent of the 571 stations throughout the nation. Most stations at that point were locally owned nonprofit outfits operated by an array of ethnic, religious, educational, and labor groups. The urban airwaves were especially filled with locally oriented broadcasts that typically catered to the panorama of interests and tastes held by well-represented immigrant groups. Through this early locally based broadcasting, immigrant communities frequently enjoyed ethnic nationality hours, locally known (often ethnic) performers, news broadcasts, and musical programs conveyed in their native tongues. Chicago's WCFL, a union-owned station, reported that its foreign language and ethnic programs steeped in local talent were among its most popular.[2]

Specially designated "International Radio Nights" introduced foreign language programming to a broader audience. These nights entailed concerted efforts by listeners of all sorts to receive programs from abroad thanks to the sheer force of the transmission signal. In those early years of radio, this unprecedented type of exploratory listening symbolized an accomplishment worthy of celebration. According to the *New York Times*, radio magazine editors and stations received scores of letters that "triumphantly" celebrated the feats of long-distance listening. The *Times*'s coverage of one such night in November 1924 approvingly documented the experiences of Dr. O.C. Risch of Passaic, New Jersey. The doctor reported hearing "a man talking in French and a woman singing in some other foreign language." Another listener kept a detailed log of reception successes that same evening. At 11:25 p.m., the listener recorded having heard a man speaking "in a foreign language, probably Italian." At precisely 11:57 p.m., this same radio enthusiast noted somewhat vaguely his approval of a broadcast that revealed "a woman talking a foreign language."[3]

This seeming enthusiasm for foreign language broadcasting in the medium's earliest years was often tied to notions of using radio to foster connections beyond the nation. Into the 1930s, for example, the stated goal of many such programs broadcast over popular American English-language stations was to foster better understanding of Latin America among English-speaking US listeners. Announcers for such programs spoke in English and Spanish (as well as Portuguese when relevant), a practice that print advertisements underscored when promoting these "special" programs for English-speaking listeners. These shows touting pan-American exchanges typically prioritized playing Spanish-language music and other highbrow cultural products, which combined to provide a romanticized view of the larger Americas beyond the United States. "The international language of music will predominate on these programs," touted one such advertisement, "but the increasing study of Spanish in the United States and English in Latin America will permit large audiences to appreciate spoken messages." Their stated purpose was to breach distances and divides across international borders. Such pronouncements harkened back to nineteenth-century notions that the transatlantic exchange of highbrow symphonic music could serve as a universal language and, in turn, help remedy societal problems. By the 1920s, such enduring notions nicely complemented the comparable contemporary beliefs in radio's value to the pursuit of such reform-minded endeavors. No longer the purview of society's elite and most educated, such internationally flavored and musically grounded broadcasts seemingly positioned ever increasing numbers of native English-speaking listeners to acquire a type of beneficial cultural capital and the ability to engage with the elite cultural products of other esteemed nations.[4]

The 1920s, however, were also a decade marked by intensifying linguistic, xenophobic, and outright racist intolerance. It was in that context that Mexican consul Leandro Gara Leal tried to frame Mexico and its heritage as an undeserving target of such disdain. To do so, he referenced those enduring elitist convictions in highbrow music's supposedly culturally edifying value. The venue was an October 1923 luncheon organized by the Civic Music and Art Association of Los Angeles, which was attended by a variety of European and Latin American diplomats and included a performance by a band from Mexico. "The music you have heard, and the dancing you have seen are not the product of jungle savages," the consul remarked in reference to that performance, "but represent the flower and fruit of one of the oldest civilizations." By then tracing his country's musical heritage to the ancient Greeks and "the music of Apollo," as well as modern Christians "who paint their angels with harps in their hand," Gara hoped to impress upon those in attendance that "music is the most powerful force for breaking down the barriers of prejudice and misunderstanding and erecting in their place a palace of fraternity and peace." It was this same rationale that infused the promotion of educational programs touting highbrow music and language appreciation. However, by framing Mexico's musical heritage as falling squarely in the European elite cultural tradition, explicitly and deliberately distancing it from the supposed musical customs of "jungle savages," Gara also implicitly acknowledged the era's increasing intolerance and racialized views directed toward Spanish-speaking Mexicans living in the United States. Mexicans were not alone in this regard. By the 1920s growing numbers of Americans viewed non–English speakers of many different backgrounds as inferior, uneducated, and unwelcome outsiders.[5]

For that reason, domestically based foreign language broadcasting that targeted foreign language speakers living within US borders was an entirely different animal from those framed as vehicles for pan-American musical and cultural exchange. Listeners to those domestic broadcasts presumably remained more comfortable in their native tongue. Such broadcasts, in turn, confronted a historical suspicion of linguistic diversity in the United States. English, as the country's national language, was more than just a shared means of communication among the majority population. As the historically spoken language of the country's Anglo-descended elites, fluency in English was essential if one aspired to political, economic, and cultural advancement. The possibility that another language might replace English as the country's primary language, however remote, threatened to undermine the English-speaking population's societal status and ambitions. This "English-centrism" and widespread conviction that monolingualism was essential to a healthy and sustainable American political culture even predates the establishment of the American nation itself. As far back as 1758, Benjamin Franklin echoed concerns of many others (including some fellow Founding Fathers) with his complaint that pre-independence Pennsylvania "will in a few Years become a German Colony: Instead of their Learning our Language we must learn their's [*sic*], or live as in a foreign country."[6]

Such suspicions expanded into a far more pervasive linguistic intolerance by the latter part of the nineteenth century. Approximately twenty-six million immigrants arrived in the United States between the 1870s and 1920s. This influx of foreign peoples, predominantly from southern and eastern Europe, coincided with a post–Civil War push for national unity. That combination triggered an unprecedented rise in linguistic intolerance and nativism that intensified into the 1920s. By that time, amidst heightened patriotism in the wake of the First World War, the popular "eugenics" movement used pseudo-scientific theories to argue that the lesser "races" of unassimilable immigrant peoples threatened the vitality of American society.[7]

The rise of eugenics, underpinned by the principles of the emergent discipline of evolutionary biology, provided a purportedly scientific justification to exclude supposedly inferior peoples from the American nation. These most recent immigrants were not just culturally different, but supposedly saddled with immutable biological traits that made them economically inefficient, intellectually deficient, vulnerable to both mental and physical diseases, and prone to poverty, dependence, and criminality. Therefore, "science" dictated that the very presence of such large numbers of inferior human beings in the United States risked undermining the country's economic progress and the durability of American society and its democracy.[8]

These trends pushed American nativism to its apex by the 1920s. Since the late nineteenth century, US immigration policy responded to these eugenic-based convictions by prioritizing the identification of and banning entry to "defective" immigrants. For any defectives that may have slipped through that net—or the native-born who displayed perceived abnormal behaviors, for that matter—an array of state-level eugenics laws allowed for institutionalization and sterilization to facilitate the end of a purportedly inferior genetic line. Such efforts to strengthen the nation through exclusion and eradication culminated with the congressional passage of the 1924 Johnson-Reed Immigration Restriction Act, the defining nativist act of this era. It set an overall limit of 150,000 immigrants a year; through low nation-specific quotas or outright bans the law virtually eliminated the entry of presumably "lesser" peoples (especially those from southern and eastern Europe, as well as Asia). This increasingly strong opposition to ethnic pluralism and immigration hardened convictions that speaking English was essential to being authentically American. The decade consequently experienced an increase of so-called 100 percent Americanization programs. Numerous "English-only" language initiatives were among the many manifestations of this surging racialized American nationalism. These trends collectively revealed the extent to which American political and popular culture had by the 1920s traded in intolerance and fears of difference.[9]

Belief in the biological immutability of such threatening differences, however, was not an uncontested position. A growing number of Americans instead perceived the problem to be culturally based, not biologically immutable, differences. Such differences were a threat, to be sure, but ones that might be mitigated through the application of broadcasting's purported transformative powers. Radio, proclaimed a regular columnist in the September 1927 issue of *Radio News*, was the "new melting pot" through which the power of Americanization and the English language would envelop the country. Addressing the era's pungent fears "that the 'Anglo-Saxon' elements in our civilization are being rapidly subordinated to the high proportion of unassimilated immigrants" and that "English as the predominant language is being displaced by alien tongues," the notably sanguine columnist Charles Adams instead assured his readers that radio was the panacea. First, he proclaimed somewhat misleadingly that a "careful search and inquiry" revealed no foreign language stations operated in the United States, evidently discounting the relevance of foreign language programs that nonetheless still aired on stations across the country. That not-so-careful search aside, Adams was confident that "the predominance of English in our spoken language will become more pronounced." Groups that "still cling to their alien tongues will have English forced upon them the more they listen to broadcasting." Radio would inevitably prove "to be an important if unconscious Americanizing influence to the extent that speaking English makes for Americanization."[10]

And yet, the highly nativist decade in which Adams wrote did indeed still see an abundance of non-English broadcasting directed at immigrant audiences. This paradox reflected the lack of established "American" radio practices. No dominant system or popular listening habits had been pervasively accepted as the "norms" against which anything atypical might be measured. The medium of broadcasting itself was entirely new and unusual. In 1922, less than 1 percent of American households owned a radio. Throughout the 1920s, a diverse array of radio interests, including immigrant, labor, educational, civic, and religious groups, vigorously contested how radio broadcasting should be developed as the medium grew in popularity. They debated over different models of station ownership, frequency allocation, programming, and regulatory structures. A significant portion of these radio activists (including many foreign language educators) vehemently opposed corporate control of American radio. Some envisioned broadcasting as a refuge from consumerism, a vehicle to celebrate cultural diversity, or a tool for high-minded education. Against such proposals stood the corporate vision of entertainment broadcasting as a national, profit-driven, advertising-supported medium that catered to a large and predominantly white middle-class audience. In its first annual report to Congress from 1927, the newly established Federal Radio Commission confirmed that the corporate model was by no means an overwhelming favorite. Summarizing listener correspondence received during its first year of existence, the FRC reported conflicting sentiments that professed "chain broadcasting is either the greatest blessing or curse of broadcasting." The FRC perhaps stated the obvious by equating a specific position with the "location of a listener or his particular taste."[11]

Amidst this lack of consensus, proponents of the corporate vision convincingly settled the matter in their favor. Prior to 1927, the United States lacked a legally constituted authority to regulate the allocation of licenses on the eighty-nine available frequencies based on stipulated standards. Consequently, the approximately 25 percent of households that owned radios in 1927 heard a cacophony of hundreds of stations, large and small, that flooded the airwaves and caused endemic interference with each other. Added into this mix were NBC and CBS, the newly established corporate radio networks with national broadcasting aspirations founded in 1926 and 1927, respectively. Both sought to secure their place on this congested frequency spectrum at the expense of smaller, more locally focused, and cash-starved competitors. With NBC and CBS taking the lead, commercial broadcasting interests capitalized on their superior unity of purpose, political connections, and financial resources to ensure their preferred broadcasting model prevailed. They overcame widespread resistance to commercialization by parlaying their inherent advantages to forge important relationships with influential legislators, officials, and regulators. These relationships, in turn, helped shape the emergence of a regulatory framework, initially through the 1927 Radio Act and corollary establishment of the FRC, which gave preference to national commercial broadcasters and advertisers in decisions pertaining to station licensing, frequency allocation, and transmission power. Regulators set technical and commercials standards for the best licenses that were primarily attainable by deep-pocketed corporate radio interests. The 1934 Communications Act, which created the Federal Communications Commission (FCC) as the successor to the FRC, reaffirmed this approach.[12]

This emergent system essentially gave the largest commercial broadcasters priority over the vast majority of the eighty-nine available radio frequencies. Moreover, that majority included the preponderance of frequencies that offered the longest transmission range and highest quality reception. One of the FRC's most consequential and enduring early actions was to designate some forty frequencies as "clear channels." A single commercial broadcaster received an exclusive national license to a specific clear channel and could broadcast at 50,000 watts, the highest power level permitted. By 1931, NBC, CBS, and their respective affiliates provided nearly 70 percent of American broadcasting and by the mid-1930s accounted for an astonishing 97 percent of all nighttime broadcasting. Along the way, receiver ownership shot up to 45 percent of households by 1930, hit 67 percent in 1935, and ultimately exceeded 80 percent by 1940.[13]

The growth that networks and their affiliates enjoyed generally came at the expense and marginalization of smaller independent stations, including foreign language broadcasters. With forty clear channel stations rendered the exclusive purview of the large commercial broadcasters and other prime spots on the frequency spectrum often licensed to network affiliates, hundreds of smaller broadcasters were left without the space to continue their operations. Those smaller broadcasters allowed to remain on the air typically received a lower class of license, which allowed their stations to operate in the shadows of the corporate behemoths. As a condition of receiving such licenses, the FRC/FCC typically assigned these smaller and technically less sophisticated broadcasters to less desirable and more interference prone locations, which were often near the extreme ends of the radio dial. The licenses required low power transmission to limit the signal's range and the potential to create interference. Furthermore, many such smaller broadcasters did not even get sole use of the frequency to which they were assigned; instead, they were forced to share their assigned frequency according to a set schedule with one or more other small radio stations. The approach maximized the number of stations that could be assigned to the limited number of remaining frequencies without undermining the commercial preferences that underpinned licensing decisions.[14]

The commercial programming that dominated the airwaves by the 1930s was both varied in its content yet narrow in its focus. Popular national shows broadcast by the ascendant networks ranged from the high culture inspired performances of the popular NBC Symphony, to the linguistic slapstick of Abbott and Costello, to the racial and ethnic stereotyping that generated huge national audiences for radio comedies such as *Amos 'n' Andy* and Eddie Cantor's *Chase and Sanborn Hour*. For all their differences, these network programs generally had one common denominator: they spoke to a restricted subset of values and presumptions that made them popular with the predominantly white middle class that made up so-called mainstream America in the 1930s. This demographic was, not coincidentally, the cross-section of the country perceived to hold the greatest aggregate purchasing power by the network stations and their advertisers, who sought the widest possible audience of potential consumers. With network stations occupying choice frequencies in terms of location and reception, increasingly regular and rigid scheduling practices made it simple for the growing number of listeners to find their favorite programs and advertisers to find their desired audience. This was no small consideration for a business model dependent on advertising revenue.[15]

Meanwhile, the number of non–English language broadcasters trended sharply downward. Three surveys of foreign language broadcasters conducted by the *Modern Language Journal* between 1931 and 1938 underscored foreign language broadcasting's diminished presence during the decade. In a sample of approximately fifty stations, the number broadcasting any form of foreign language programming declined from thirty-two to fourteen during those years. By 1940 the FCC identified 199 stations out of the approximately 850 licensed stations—23 percent—broadcasting some sort of foreign language programming. Any impressiveness of that 23 percent figure is mitigated by three corollary findings: first, the majority of those stations operated on 250 watts of power or less, far below the average power of a network station; second, the FCC ascertained that some fifty-seven stations had entirely dropped foreign language programming from their schedules shortly before the survey; third, those 199 stations combined for a rather paltry 1,330 hours of weekly programming throughout the entire country.[16]

The qualitative evidence indicating commercialization's significant impact on foreign language broadcasting is also considerable and convincing. Labor-oriented stations like New York's WEVD and Chicago's WCFL were not exclusively foreign language broadcasters; their varied offerings, however, included foreign language programming directed toward their large working-class immigrant audiences. Regulators forced both stations to reduce their power and move to less desirable shared frequencies. Historian Derek Vaillant documents the regulatory hostility that the FRC directed at Chicago's once vibrant foreign language broadcasting market during the late 1920s. In response to its actions, the FRC received a multitude of letters from angry listeners unwilling to accept the favoritism that agency exhibited toward networks and the agency's efforts to, in Vaillant's words, "relegate foreign language and independent broadcasters to the musical margins of broadcasting." In 1934, two Brooklyn stations faced the revocation of their respective licenses due to their tendency to give, as the official recommendation stated, "excessive time to foreign language programming."[17]

Foreign language broadcaster WSAR of Fall River, Massachusetts, scrambled to adjust to these adverse conditions. This small station kept its license. It was even permitted to increase its power from 250 watts to a still-modest 1,000 watts in 1938. The FCC nonetheless pushed it to the high end of the AM radio dial. It continued to broadcast foreign language entertainment programming from its marginal perch on the frequency spectrum into the late 1930s. However, in recognition that its English-speaking audience had long since embraced other programming preferences, WSAR made a conscious decision to eliminate its foreign language educational programming, the type of non-English programming most likely to attract an English-speaking middle-class American audience. "Listeners were likely to treat these [educational] programs as novelties, and the programs did not attract serious interest," the station manager remarked in 1938. WSAR instead concentrated on reaching an immigrant audience through foreign language entertainment.[18]

WSAR's reputation as source of news and entertainment for foreign-born listeners was evidently well established by the end of 1939. In December of that year, thirty-year-old Louis Bielecki stormed into WSAR's studios. He interrupted the "Portuguese Hour" live musical program to protest the joint German and Soviet invasion of his native Poland just months earlier, which triggered the start of the Second World War in Europe. The Portuguese entertainers were silenced mid-verse and remained seated in their chairs, "cowed in fear of being shot if they moved," as a front-page *Boston Globe* story put it. They had good reason. Bielecki held what appeared to be a gun concealed within his jacket pocket. He grabbed the live microphone, and proclaimed, "I'm going to broadcast for 10 or 15 minutes in the interests of Poland and anyone moving or attempting to stop me will be shot." The "staunch Polish patriot," as *Radio-Craft* magazine described him, then delivered an anti-Nazi harangue for more than twenty minutes in the hopes of mobilizing wider Polish immigrant support for his distant homeland.[19]

In fact, only the first two minutes of Bielecki's tirade went out over the air. The program director managed to mute the microphone when the intruder was momentarily distracted; the station's sports director Orville Seagrave had unexpectedly burst into the studio. Seeking the reason for why a "strange voice in a strange tongue" had suddenly preempted the station's regularly scheduled program, he was instead told to take a seat with the rest of the hostages. Ultimately, station personnel were able to summon the police, who arrived and quickly realized that Bielecki's supposed gun was nothing more than a wooden mallet. Bielecki was evidently a creature of habit. Some seven years earlier, WSAR's would-be hijacker had been arrested with his two brothers on a charge of vagrancy after creating a public disturbance. Bielecki, whose preference for the mallet apparently long predated his ill-fated commandeering of WSAR, was given an additional charge of carrying a wooden weapon and sentenced to six months for this 1932 incident. Though the hot-tempered rabble rouser's luck evidently had not changed much in the subsequent seven years, *Radio-Craft*'s account of the incident reported that "Mr. B. had done his bit for the motherland, his conscience was relieved." The botched seizure of WSAR notwithstanding, that Bielecki saw the station as the ideal vehicle to reach a small but nonetheless like-minded audience of fellow patriotic Polish-speaking immigrants speaks to the programming adjustments that WSAR made by the eve of American entry into World War II. By then WSAR was a low power station located at the fringe of the radio dial largely dependent on a relatively small audience of immigrant listeners living in its proximate area. WSAR thus exemplified the limits foreign language broadcasters faced by the 1930s in terms of frequency assignments, transmission power, and audience demand.[20]

The once heralded international nights could not fare much better in this climate. Throughout the world, governments sought to impose order and control over increasingly congested airwaves. In the United States and elsewhere, international agreements complemented domestic regulatory policies to allocate frequencies and set technical standards in ways that sought to minimize the intrusion of foreign-generated interference on domestic programming. As the thrill of hearing transmissions from a distant nation lost its novelty in the late 1920s, an ever-expanding pool of American listeners increasingly demanded static-free reception of the major American stations and their most popular programs.[21]

As the presence of foreign languages generally decreased from the mainstream American airwaves, spoken Spanish had become especially disdained. Many Americans viewed Spanish-speaking immigrants from Latin America, especially Mexico, as an unassimilable and racially inferior "other." In the process their spoken language became a marker of that inferiority. The Spanish language was essentially a victim of the dominant racial discourse that had emerged since the nineteenth century; it was a discourse that castigated Mexicans, as well as other Latin Americans judged guilty by association, as an inferior "mongrel" mix of Spanish and Indigenous blood. An 1896 federal court decision that legally defined Mexicans as "white" (popular perceptions notwithstanding) further fueled these fears. That legal designation made Mexicans, unlike Asians, eligible for citizenship under US naturalization laws. Then, in the 1920s, just as American nativism was hitting its crescendo, the number of Mexicans entering the United States spiked upward amidst the enduring violence of the Mexican Revolution. The 1924 Johnson-Reed Immigration Restriction Act could not help stem this tide, since the law deliberately exempted Mexicans from any limits at the behest of an agricultural lobby seeking to protect a cheap source of foreign farm labor. Fears abounded that the increasing presence of citizenship-eligible inferior Mexicans and other Latin Americans in the United States risked an internal weakening of the American nation itself. In this context, spoken Spanish served as an audible and racialized reminder of that demographic's unwelcome, threatening but still growing presence in the United States.[22]

Not surprisingly, amidst this spiraling hostility the Spanish language became increasingly unwelcome on the airwaves. In California, home to a particularly large population of first- and second-generation Spanish-speaking Mexicans, officials sought, albeit unsuccessfully, to ban Spanish entirely from the airwaves. Though pushing for an official ban was an overreach, the hostile climate of the 1930s saw advertising-minded stations in California and Texas (the latter also with a large Spanish-speaking Mexican population) voluntarily decrease such programming while leaving what remained at undesirable listening hours unlikely to generate much of an audience or ad revenue.[23]

Broadcasters based in Mexico that targeted the US market were forced to pivot as well. Norman Baker was a quack doctor and charlatan broadcaster forced off the air by American authorities, but invited by Mexico to broadcast into the United States from the border city of Nuevo Laredo. Mexico was using Baker and other "border blaster" broadcasters to disrupt American radio. The goal was to pressure the United States to relinquish its stranglehold over the hemisphere's available radio frequencies (the topic of the next chapter). However, Mexico also forced Baker to abide by its radio rules that limited the number of English-language broadcasts within a twenty-four-hour period and stipulated that advertisements first needed to be spoken in Spanish before English. Baker balked. "When the English language audience hears Spanish, they turn their dial and seek another station," Baker protested to the Mexican authorities in 1936. Meanwhile, other bona fide Mexican broadcasters who used their stations to attract American tourism to their locales sought to avoid any trace of a Spanish-language accent in the English language portions of the programs so as not to alienate any potential travelers.[24]

The impact of commercialization and regulation on foreign language broadcasting extended beyond the resulting downward trends in frequency assignments, transmission power, audience numbers, and actual broadcasts. It defined the very meaning of "American" radio. Enjoying political clout and a substantive presence on the airwaves by the early 1930s, corporate broadcasters found themselves in a particularly strong position to present their activities as the definitive "American" plan for radio offering decidedly "American" programming. Enjoying easy access to both government regulators and a nationwide audience, commercial broadcasters presented the system they dominated as the only logical structure for American radio. Corporate radio interests, in concert with the FRC and then the FCC, lambasted proposals for alternative broadcasting models as catering to "special interests" often intent on broadcasting narrow-minded "propaganda." Such ill-advised efforts to fill the public airwaves with programs of little use to most Americans, they argued, were not consistent with American freedom and democracy. Commercial network broadcasters insisted their programs, which sought to reach and engage the largest possible national audience, far more effectively promoted broadly based American ideals. Even the anti-corporate climate of the Depression years could not stem the tide. Advertising-based national radio seemed to be one of the few industries that made a positive contribution to the struggling American economy during an otherwise dark period. Consequently, many Americans proved receptive to this definition of an "American system" of radio.[25]

Dr. A. Holmes Johnson typifies the listener who came to equate commercial network broadcasting as distinctly American. In 1939, this one-time Oregonian who had moved to distant Kodiak, Alaska, wrote a letter of thanks to the FCC. He was grateful it licensed a new shortwave station that carried NBC programs all the way to his remote outpost. Johnson did, however, have two complaints. First, NBC program schedules meshed awkwardly with Alaska's time zone, forcing the audience to listen to popular programs at inconvenient times. Second, Johnson implored the FCC to license a similar station to carry CBS programs. If the FCC would help Alaskans obtain CBS programming alongside NBC's, "we would be in almost as good a position as if we were in the States."[26]

Johnson's musings illustrate how the purported American system of broadcasting engaged American national identity. Political scientist and identity scholar Benedict Anderson locates the origins of this media–national identity nexus in early modern Europe. A push for ever-larger markets on the continent facilitated the absorption of multiple local dialects into standardized national languages generally identified with specific geographies and ruling regimes. This transformation facilitated the growth of mass publications, such as books and newspapers, that relied on a standardized national language to attract the largest possible audience. Readers of these publications became cognizant of millions more anonymous yet common language users reading the same material but who resided well beyond one's locality. This emergent sense of a connection and attachment to a larger nation captured these readers' realizations that they were part of an "imagined community" of thousands, even millions, of people who they would never meet but who nonetheless presumably shared many of the same national values and experiences. Network radio served a comparable purpose for Johnson. He imagined a connection to the millions of other anonymous yet nationally linked American listeners that he knew shared his programming preferences and, presumably, his values from thousands of miles away.[27]

To be sure, these conclusions are not meant to exaggerate radio's oft-presumed power to unify an infinite number of diverse peoples into a monolithic culturally unified body. For one, the so-called American system was not nearly as all-compassing and homogenous as the national radio proponents would have had everyone believe. Significant swaths of the country far less remote than Alaska had not been enveloped into the network system. Those geographic expanses were primarily in the South and the West, and also included many areas with low populations or other audience demographics deemed to be unprofitable for national radio. Decisions to ignore supposedly unprofitable markets left space for a complicated array of local, regional, and even international broadcasters to fill in the gaps. They often did so through programming that did not fit neatly into the American system as defined by the national networks. Such "niche" programming, including the ethnic and foreign language programs found on many lower power stations, were among the programs that filled in the gaps and that proved very important to the communities they served. This programming could simultaneously engage audiences that were not well served by the dominant "American" system while concurrently provoking the hostility of others who deemed such seemingly non-normative programs to be an affront to commercialized and network dominated broadcasting that purportedly represented the American system.[28]

The solidifying of this commercialized notion of American radio in the 1930s thereby intersected with the linguistic intolerance that pervaded American society by the twentieth century. Unlike the experimental broadcasting in the early 1920s, by the 1930s popular understandings and practices of American broadcasting had become clearly tethered to the outlooks, values, and presumptions of a large, historically privileged, and predominantly white middle-class audience. This mainstream audience's sense of their "American" identity had historical roots in the shared use of the English language. The interconnected understandings of their American identity and the programming deemed acceptable for American radio proved a potent mix. From that vantage point, the seemingly indecipherable ramblings of foreign language broadcasts emerging from the loudspeaker now challenged many average listeners' sense of their identity, place, and history. They were prepared to respond.

The Imagined Menace of Foreign Language Broadcasting at Home

Kingston, Pennsylvania, resident Earl Andrews, whose rant against foreign language broadcasts as decidedly un-American opened this chapter, was far from alone in his sentiments. T.L. McCarthy of Yonkers, New York, and Brooklyn resident Sylvester Sullivan echoed Andrews's intolerance. "This is America," McCarthy declared to the FCC in 1935, "and there should not be allowed any tongue to speak any but the English language over the radio." Sullivan's letter, written with similar hostility, contained a laundry list of foreign languages he found offensive when broadcast over the airwaves. His catalog of taboo languages included German and French, as well as Yiddish and "Hindustance." Sullivan wondered why he had to hear such drivel "every time" he turned on the radio when he was "in America, the official language of which is English."[29]

All three letters ascribed to English an elevated status as the country's presumed national language. By extension, Andrews, McCarthy, and Sullivan also attributed an elevated status to themselves vis-à-vis immigrant groups by virtue of their own fluency in the national language. To be sure, English lacked legal standing as the "official" language of the United States, Sullivan's claim to the contrary notwithstanding. In fact, a spate of Supreme Court rulings in the 1920s overturned a handful of nativist-inspired state efforts to provide English with such a formal status. And yet, even President Franklin Roosevelt, a Democrat with strong support among immigrant groups, repeated the incorrect claim that English was the United States' "official language." He did so in 1937 while lamenting that after four decades of American control over Puerto Rico so few residents of the Spanish-speaking colony spoke or understood English. Earl Andrews's harangue, in which he placed himself on a plane with great American presidents who, like him (obviously) spoke English, effectively diminished non-English-speaking immigrants by ascribing to them an inferior status based on their purported lack of English knowledge. McCarthy's call to ban all non-English broadcasts and Sullivan's mistaken claim that English was his country's "official" language served a similar purpose. Immigrant groups could never aspire to reach a comparable stature so long as they clung to distasteful, un-American foreign tongues.[30]

From Hugh Milliken's standpoint, it was obvious that commercial radio was synonymous with American radio. As such, all commercial radio, this New York City resident believed, should be in English. "If advertisers' clientele can't speak English," he wrote to the FCC in 1934, "they should—so should the broadcasters and the sooner the better."[31] One might have argued that immigrant groups embracing non-English but nonetheless American style advertising-supported commercial broadcasting had, in fact, taken one step in a longer process of assimilation into American society. For Milliken, though, there was no room for such nuance. Foreign language commercial broadcasting simply corrupted American radio. Commercial broadcasting in the United States had to be in English. Anything less was un-American and unacceptable.

The linguistic intolerance manifested by William Seeley Logan spoke to concerns about immigrant assimilation. Logan, who was blind, expressed his irritation at encountering foreign language programming while tuning the radio to his favorite shows. "As this is America," the Arlington, New Jersey, resident complained in his 1934 letter to the FCC, "why is permission given to broadcast in foreign languages?" Channeling Benjamin Franklin's anxieties about the Germans in Pennsylvania from almost two hundred years earlier, Logan asked, "We want these people to live among us but why don't they learn our language?" Logan believed that radio's presumed potential to educate, unite, and uplift should be used to school the large immigrant population rather than cater to their foreign tastes. "I believe the Radio has an important position in our Educational development," Logan concluded. In his view, foreign language broadcasting did not harness radio's full educational potential to assimilate its immigrant audience.[32]

Logan's attention to assimilation reflected a more nuanced linguistic intolerance than the simplistic one-dimensional "English-only" grievances levied by the other complainants. By presuming that radio was an underutilized tool of assimilation but with potentially harmful consequences for the immigrant audience, Logan's views reflected the subtle influence of the increasingly credible social science disciplines like anthropology by the 1930s. The social science disciplines that emerged by the early twentieth century had started to undermine, albeit sometimes imperfectly and inconsistently, the older racially rigid eugenic theories of difference that seemingly peaked in their popularity just one decade earlier. These new intellectual disciplines began to increasingly emphasize cultural and environmental explanations of human diversity, which theoretically could be bridged through education and experience. The corresponding rise of a more inclusive "civic" nationalism in American society welcoming any worthy individual as a valuable member of an American nation—including "Americanized" immigrants—underscores the increasing strength of these influential disciplines. Mass media occupied a central role in these changing perceptions. Progressive-minded reformers had first challenged nativist and eugenicist-inspired opposition to immigration by heralding the Americanizing power of movies to assimilate and educate the foreign-born population. Not surprisingly, that earlier tendency to view movies through the prism of their assimilationist potential coincided with the popular acceptance of an American model of cinema that also catered to middle-class ideals and aspirations.[33]

The application of this same assimilationist presumption to commercial broadcasting in the 1930s came at a unique moment for immigrants. By that time, many in the final wave of immigrant Americans who had arrived prior to the 1924 passage of the Johnson-Reed Immigration Restriction Act had gained their citizenship and could vote for the first time. President Franklin D. Roosevelt, a Democrat who refused to scapegoat foreign-born peoples for the

Depression, captured a large majority of these voters. Roosevelt also publicly celebrated the strengths the United States enjoyed from its foreign-born but Americanized citizens. (The notable exception here was Asian Americans, especially Japanese Americans. Roosevelt ordered Japanese Americans, foreign- and native-born alike, incarcerated during World War II, which reflected widespread fears that this racially distinct and devious people could never truly Americanize.)[34] Logan's position was consistent with this broader view that some immigrant groups might not yet count as "real" Americans; however, through proper education and cultivation their potential for Americanization and advancement was wide open, at least for those with European roots.

K. St. Clair's opposition to foreign language broadcasting captured another increasingly popular assimilationist outlook. The Fort Hamilton, New York, resident professed in 1934 that foreign *oriented* broadcasts were "perfectly all-right." Foreign *language* broadcasts, however, were not. To St. Clair, spoken English was a primary benchmark of assimilation, through which expressions of ethnic diversity could be rendered acceptable. Programs catering to the European tastes of an immigrant group, St. Clair proclaimed, "should be carried on strictly in the English language. It is annoying to tune in on some station and hear alot of gibberish which one cannot understand coming out of a loud-speaker."[35]

St. Clair's views appear paradoxical: a linguistically intolerant embrace of ethnic diversity. If the emergence of the social sciences by the early twentieth century contributed to a wider appreciation of human differences by St. Clair's time, this gradual and uneven transformation of popular attitudes was not necessarily incompatible with deeply rooted linguistic intolerance. Perhaps St. Clair's aversion to languages that signified particularistic ethnic identities raises questions about his genuine commitment to ethnic diversity. Nonetheless, his willingness to accept English language programs that appealed to immigrant interests and cultural values reflects a measured, if limited, acceptance of ethnic pluralism that had had little traction only a decade earlier. Like Logan, St. Clair saw in radio a means of managing diversity in ways that made it more acceptable to mainstream America.

By the late 1930s, with war looming over both Asia and Europe, concerns about foreign language radio took on a new resonance beyond the historic linguistic intolerance. Many Americans now also feared that foreign language radio could be propagandizing among potentially disloyal immigrant populations, which threatened to undermine American neutrality and democracy.[36] A resurgence of anti-immigrant "racialized" nationalism during this period helped fuel such fears. Increasingly pervasive impulses to blame inherently inassimilable and disruptive immigrants for lingering American economic difficulties accompanied growing conservative opposition to Roosevelt and his New Deal economic policies.[37] William Bauer's 1937 letter to the *New York Times* captured those sentiments. He identified foreign language radio as a key vehicle used by "certain foreign groups" seeking to "consolidate alien nationals into groups hostile to our form of government and nation." According to Bauer, these un-American interlopers used "radio propaganda chiefly in foreign languages" to collect money "for the support of foreign powers." "I am not a xenophobe," Bauer earnestly but perhaps unconvincingly claimed. He implored "every intelligent citizen" to censure current and past presidential administrations for allowing "certain alien groups to indulge in activities that openly insult our nation and foment ill-feelings that are contrary to the spirit of American democracy and endanger our national unity."[38]

Bauer was not alone in his anxieties. With both Asia and Europe at war by 1940, concerns about and opposition to foreign language broadcasting continued to grow. New York radio programming directors expressed a consensus opinion that "the public is fed up with foreign language speakers unless they have a mighty important message." Colonel Walter Bell from Danbury, Connecticut, expressed concern that the local, state, and federal governments had failed to install adequate protections to ensure that "alien language" broadcasting did not provoke "fifth column" activities. Bell urged the proper authorities to "lock the stable while the horse is still in there" and "issue orders that such broadcasting in the language of the enemies of civilization be stopped." New York City's Walter Grove shared Bell's fears that foreign language broadcasting threatened to provoke the foreign-born into committing internal sabotage against the United States. Underscoring foreign language radio's marginalization, Grove railed against "some of the smaller stations on the upper third of the kilocycle band used in America" for broadcasting propaganda in support of fascist Italy. With apocalyptic urgency, he warned that "the American people will have to take these things in hand immediately if they do not wish liberty to be an academic subject behind locked windows." Failing to act would only postpone the inevitable "day of reckoning." A CBS affiliate in Rhode Island, mindful of such sentiments, canceled its two French and Italian language programs on international affairs in June 1940. That was the month that Italy joined Germany and declared war on both France and Britain, followed soon after by France's surrender to Germany and the installation of France's pro-Nazi collaborationist "Vichy" regime.[39]

Even the renowned comedian Groucho Marx, a giant in the history of American entertainment, felt aggrieved by foreign language radio. "According to all reports, there are thousands of phony and disloyal Americans living in the United States," he groaned. "If this is true, it might be a good idea to forbid the broadcasting of foreign language programs."[40] His grievance stood apart from the contrarian and subversive public persona he cultivated. This immensely talented comedian enjoyed a career on stage, film, radio, and television that spanned the better part of the twentieth century, most notably as a member of the Marx Brothers, whose popularity peaked in the 1930s. During the height of his fame before the Second World War, he also made occasional contributions as a humorist to major newspapers and magazines. In one such nationally syndicated essay entitled "What This Country Needs," Marx joked that he would run for vice president in the upcoming 1940 election. In nearly two thousand words published that June, he offered a political platform that envisioned a country with good ham sandwiches and attractive corset-wearing women. He called for a traffic-free world where people traveled on oxen. He demanded vacuum cleaners that "won't scare the daylights out of you by whining like a Boeing bomber" and envisioned a "national school for lovemaking." The subversiveness of Marx's public persona ultimately compelled the FBI to investigate him in the early 1950s amidst the intensifying Cold War for his alleged pro-communist and pro-Soviet sympathies dating back to the mid-1930s.[41]

Figure 1. Groucho Marx is second up from the bottom wearing glasses with a cigar. 1931. Courtesy of the New York-Telegram and Sun Newspaper Photograph Collection, Library of Congress Prints and Photographs Division.

But when the subject shifted to foreign language broadcasting in the United States, it was a very serious Marx who feared subversion. He articulated his concerns about foreign language radio comments not in his capacity as a comedian or a columnist, but as a concerned citizen. They appeared in a curt letter to the editor published in the August 30, 1940, edition of the *Los Angeles Times.* "The air is full of them," Marx railed with obvious disdain toward foreign language radio programs. "I'm sure they would not be permitted or tolerated in any other country." In all, he expressed his views in fewer than sixty joke-free words.[42]

Assimilationist notions of what made a proper American informed Marx's own vehement opposition to foreign language radio articulated in 1940. He was a child of immigrants. The comedian's non-English-speaking grandparents, entertainers themselves, received the brunt of his personal contempt for their own lack of assimilation into American society. Referring to their decision to migrate to America in the late nineteenth century only to find their show business careers stagnate, a sarcastic Marx wrote that "for some reason there seemed to be practically no demand for a German ventriloquist and a woman harpist who yodeled in a foreign language." After other occupational setbacks, Marx derisively noted that his German-speaking grandfather settled into a new career that "consisted of never doing another day's work until he died, forty-nine years later."[43] Marx's animosity toward foreign language broadcasting thereby had a personal resonance that intersected with the broader historical notions of language, assimilation, and identity in American society.

The editors treated Groucho Marx's letter as any other. His celebrated name was not written in large bold type or accompanied by a picture of his familiar bespectacled smiling face with its grease-painted mustache. The only additional identifier accompanying his name was the customary listing of the author's locale—the Hollywood suburb of Culver City, in this case—to give readers a sense of the geographic base of a usually unknown writer. The letter itself was buried in small print in the middle of the opinion page. Only a thorough reader of the editorial section would have stumbled across it. Perhaps the pedestrian treatment accorded Groucho Marx's letter reflected the unremarkable mainstream attitude it expressed. Although the wars being fought in Europe and Asia might have served as the more immediate provocation for the entertainer, the substance of Marx's allegations against foreign language broadcasting echoed those made throughout the 1930s by virtual unknowns such as Earl Andrews, William Seely Logan, and K. St. Clair.

Two weeks later, fellow Southern Californian Robert Capps, an amateur radio operator himself, used the *Los Angeles Times* editorial page to explicitly second Groucho Marx's position. Capps urged all "loyal Americans" to endorse Marx's call to ban all foreign language broadcasting. He cited the possible "sinister purposes" of such broadcasts in an international environment plagued by foreign nations at war. Just like other objectors, Capps did not understand why foreign language broadcasting was ever allowed in the first place. Such programs, he claimed, "can benefit so few users" and are "in part or entirely not capable of being understood by the average American audience."[44] Like many of his contemporaries, Capps contrasted his understanding of genuine American broadcasting against the presumably dangerous corruption of the medium through foreign languages. Authentic American radio, Capps believed, had to supplant un-American foreign language programming and those who listened to it.

The National Council on Freedom from Censorship, an affiliate of the American Civil Liberties Union, proposed a strategy to contain the supposedly imminent threat posed by foreign language broadcasting. It petitioned the FCC in August 1940 to require that all radio stations record their non-English programming for later investigation. By calling on the broadcasters themselves to keep comprehensive records, the proposal advocated an expansion of existing FCC surveillance programs. For its part, the FCC had concerns about foreign language broadcasting, but was not wholly opposed to it. To be licensed, foreign language radio stations, the FCC maintained, should facilitate the Americanization of immigrants. Amidst ongoing wars abroad and consequently increasing fears that foreign language radio might be harnessed to mobilize unfriendly immigrant sentiments, the FCC's National Defense Plan, announced a month earlier, promised additional radio "patrolmen." They would, among other duties, record and translate foreign language broadcasts. These efforts accompanied previously announced FCC initiatives to clamp down on amateur and commercial licenses requested by noncitizens. The American Legion was not comforted by these measures. The politically powerful and influential veterans' organization voiced its objections to foreign language broadcasting that same summer. Founded after the First World War and claiming a membership that exceeded one million by 1940, the Legion had expressed concerns over disloyalty and radicalism among immigrant groups since its inception. Consistent with that tradition, the Legion passed a resolution at its 1941 convention in Milwaukee that specifically condemned foreign language broadcasting in the United States as "un-American" and "detrimental to the cause of national unity."[45]

The irony is that the presence of such potentially sinister foreign peoples in the United States had reached a nadir by 1940. With the 1924 Johnson-Reed Immigration Restriction Act having nearly eliminated new immigrant arrivals (with Mexicans being a notable exception), the percentage of the foreign-born US population steadily dropped from 13 percent of the total population in 1920, to about 11.5 percent in 1930, and down to only 8.8 percent in 1940. Moreover, the vast majority of those foreign-born had earned their citizenship by 1940. The percentage of the total foreign-born US population still without citizenship in 1940 was approximately 3 percent (having dropped from about 6.5 percent prior to the passage of the 1924 Johnson-Reed Immigration Restriction Act).[46] Moreover, just as immigration had dramatically limited the presence of foreign-born peoples in the United States, radio regulation had drastically reduced the amount of programming in foreign languages by 1940. If foreign-born peoples listening to foreign language radio could indeed pose a grave threat to American society, the statistical patterns evident during the 1930s and into the 1940s suggest that Groucho Marx and like-minded radio listeners should have been greatly comforted, not alarmed, by the state of American radio in 1940. Both foreign-born peoples and foreign language radio in the United States had declined precipitously since the 1920s. Simply put, fears expressed over the dangers of foreign language broadcasting were disproportionate to the immigration and regulatory trends that affected foreign language radio by the eve of the US entry into the Second World War.

* * *

By the early 1940s, the dominant American broadcasting practices had marginalized and obscured foreign language broadcasting, but never fully eliminated it. As foreign language broadcasting's presence on the American airwaves declined, its apparent contrast with increasingly popular notions concerning what entailed proper American radio became amplified. This process, in turn, provided the basis of attacks on foreign language broadcasting and its supposedly non-English-speaking listeners. By the 1930s, programs in languages other than English, minimal as they had become, convinced a cohort of angry listeners ranging from the virtually unknown to the nationally famous that American society faced a grave threat from within. Such overexaggerated convictions reflected the relationship that developed between the English language, American identity, and commercialized American broadcasting. The belief that foreign language radio broadcasts represented a substantive foreign intrusion into the United States also exemplified how listeners viewed the invisible "ether" as a tangible national resource intertwined with the physical territory across which broadcasts traversed within the nation's boundaries. As such, the ether was no less worthy than the nation's physical territory of being defended from foreign or otherwise malfeasant threats.

To be sure, opposition to foreign language broadcasting was not necessarily a consensus position in 1930s America. The significance of this opposition lies in what it says about how listeners engaged with broadcasting media on a broader level. In using foreign language broadcasting as the vehicle to articulate a sense of an American national identity, these English speakers conjured an image of a faceless and monolithic foreign language–speaking immigrant audience poised to undermine the nation. The aggrieved knew little, if anything at all, about the audiences they castigated. They reached their conclusions after hearing broadcasts they almost certainly did not understand. Nonetheless, these self-professed American opponents of foreign language radio sought to stifle the seemingly evident threat they claimed these broadcasts and their target audiences manifested.

At the root of these reactions lie exaggerated notions of radio's transformative powers. A radio-inspired wholescale undermining of the American nation from within by foreign language–speaking peoples was unlikely. By 1940, foreign language broadcasting had already been reduced to a minor presence within the broader canvas of US radio. Nonetheless, predisposed to linguistic intolerance and convinced of radio's immense power, the objectors had let their imaginations run wild upon hearing languages they did not speak. They imagined foreign hordes gathering from within the nation and poised to use the otherwise beneficial power of radio to run roughshod over the United States and the entire American way of life. The likes of Earl Andrews, William Seeley Logan, Robert Capps, Groucho Marx, and others believed they had, as Americans, the authority, status, and legitimacy to complain about the "gibberish" they heard on the air. In reality, the commercialization of American radio had largely accomplished what foreign language broadcasting's opponents demanded: the vast diminishment and marginalization of presumably threatening foreign language transgressions on the American airwaves.

A minor presence, however, is not synonymous with an insignificant one. In fact, as historian Ari Kelman demonstrates, the presence of multilingual and multigenerational Jewish audiences attuned to Yiddish radio illustrates that foreign language radio was not inherently working at cross-purposes with immigrants learning English or Americanizing, but instead helped facilitate it.[47] Moreover, in conjunction with radio, influence flowed in both directions. So-called ethnic influences on "mainstream" American music, cuisine, art, and other cultural forms have continued throughout American history.[48] Such complexities, however, are obscured by the persistent inclination to prioritize spoken English as a key metric to define Americanness. Ultimately, such convictions have not derailed these multigenerational processes of cultural change, exchange, and multifaceted assimilation.

More recently, inclinations to profess greater appreciation for cultural diversity and inclusion have not shaken English's privileged position as a key metric of American identity. Nor have such sentiments prevented a countervailing backlash rebuking diversity and inclusion initiatives as being decidedly un-American. Spanish, in particular, has continued to provoke a heightened level of antagonism in these enduring battles over culture and language.[49] The common denominator remains historically rooted understandings of the relationship between the English language and American identity. In this vein, Groucho Marx's intolerance for foreign language broadcasting so many decades ago speaks to that broader historical continuity. His comedy was popular across several generations of twentieth-century Americans because the talented entertainer knew how to engage the existing sensibilities of his mainstream white middle-class audience. That task was made far easier when someone like Groucho Marx shared many of those sensibilities as well.

CHAPTER 2

"An Invasion by Radio Is Crossing the Mexican Border"

John Brinkley, Border Blasters, and the Geography of American National Identity in the 1930s

It was January 1932. G.W. Coffman of Hutchinson, Kansas, and Harvey Averill of Vinita, Oklahoma, had run out of patience. Since October 1931, severe interference from a radio station in Mexico had made it nearly impossible to enjoy their favorite programs. Coffman worked for International Harvester and in his spare time played the clarinet in Hutchison's municipal band. Averill was the publisher of the *Craig County Democrat*, which he had founded just two years earlier. Their grievance was shared by many Americans across the country. Too often when these two men and countless others turned on their radios, they found that the medical quack John Brinkley had hijacked the airwaves. After the Kansas-based doctor and broadcaster lost both his US radio and Kansas medical licenses in 1930, he headed south and secured permission from Mexico to establish a powerful new station, XER, in the border city of Villa Acuña. It was located just south of the Rio Grande, opposite Del Rio, Texas, where Brinkley opened a new hospital thanks to that state's laxer medical licensing standards. Both Coffman and Averill wrote to Senator Clarence Dill, a Democrat representing Washington who was publicly pushing the State Department to do more to drive Brinkley from the airwaves. Averill complained to Dill that Brinkley "virtually ruined radioing for me." Coffman concurred. "Something should be done, and done quick," he insisted.[1]

Brinkley, to be sure, also had legions of devoted fans. Their stories have been told elsewhere.[2] This chapter instead focuses on the opposition XER's broadcasts provoked. It is a story that culminates in the 1941 enactment of the

Figure 2. Dr. John R. Brinkley. Courtesy of the George Grantham Bain Collection, Library of Congress Prints and Photographs Division.

first North American Regional Broadcasting Agreement (NARBA). That treaty resolved a decade-long US-Mexico radio dispute into which Brinkley had been entangled. The fact that Brinkley's broadcasts from Mexico ranged freely into the United States provided the focal point of the controversy. The anti-Brinkley protest letters illustrate how internalized understandings of American national identity, geography, and sovereignty incited the animosity of these listeners. Such understandings became entangled with deeply rooted "us" against "them" assumptions about the US-Mexican border's significance. It was perceived as a jurisdictional barrier intended to keep the allegedly negative and corrupting influences of Mexico from intruding into the United States. The border's evident porousness, however, often worked at cross-purposes with that goal. Outraged listeners underscored their American bona fides to justify their anger and demand that their government throttle the charlatan Brinkley's radio-driven violations of American territory and sovereignty. To make their case against XER, they deployed the tools of national identity—including their knowledge of geography, maps, and other forms of cultural capital—to demand action. That body of knowledge shaped how these infuriated listeners engaged with and reacted to the broadcasts heard over XER.[3]

Brinkley lost his KFKB license in 1930 precisely because he fell outside the perceived mainstream of acceptable American radio. Grievances came from listeners who embraced the so-called American system of commercialized broadcasting that was firmly entrenched by the 1930s. US diplomats tasked with resolving the Brinkley-driven radio dispute proved most skilled at making miscalculations and missteps. These shortcomings ensured this US-Mexico conflict would extend into the early 1940s. This chapter's analysis of anti-Brinkley opposition that spanned a decade demonstrates how the act of listening can elicit public declarations of allegiance to the nation, even in the absence of any physical threat or danger. Passions ran high against a perceived foreign intrusion that could be neither seen nor touched, an intrusion that generally posed no substantive threat to the life or livelihood of any listeners who objected. And yet, in the minds of these antagonized listeners, Brinkley had to be stopped.

A "Grudge Station" in Mexico

A typical John Brinkley broadcast over XER from November 1931 provides a sampling of the content that G.W. Coffman, Harvey Averill, and thousands more found so dreadful. A concerned woman from Aurora, Nebraska, had written Brinkley about an unknown illness that afflicted her husband. The doctor minced no words in his over-the-air response: "The gentleman has [an] infected prostate and infected glands." He then urged this woman to get immediate care for her husband and cautioned that "many men go insane because of this neglected, untreated prostate." Her husband, he warned, could become one of the many who wind up "in the insane asylum because of infection in the prostate." The only solution, the doctor insisted, was for the afflicted husband to make the journey to his hospital, at that point still located in Milford, Kansas, and which happened to specialize in prostates.[4]

If this afflicted man managed to make the journey, he likely left the Brinkley hospital with a pair of goat testicles transplanted into his scrotum. That was Brinkley's signature operation. He developed the procedure in the 1920s and claimed it cured male impotence. As Brinkley tells it, he acted on a hunch after observing two randy goats mating while in the company of a patient suffering from impotence. After his first testicle transplant patient survived the procedure and allegedly fathered a child nine months later, Brinkley burnished the legend of his "goat gland" operation. By the 1930s, he insisted the procedure could cure nearly thirty afflictions and diseases, including diabetes, high blood pressure, and senility. His treatments were not just limited to men. He also claimed that transplanting goat glands into a woman's "sex glands" could cure female infertility, dementia, and obesity.[5]

Beyond prostates, Brinkley was not shy about discussing other discomfiting conditions over the radio. Acne, constipation, gas, and hemorrhoids were among those that also earned Brinkley's attention. His proposed treatments ranged from merely phony to outright dangerous. He also used his radio program to perpetrate a type of pharmaceutical fraud. He "prescribed" medications over the air based on the descriptions of symptoms outlined in the letters he received. These prescriptions, however, were nothing more than over-the-counter medications repackaged as Brinkley medicines and sold by cooperating pharmacists at a tremendous mark-up. Both Brinkley and the pharmacists profited handsomely from the scam. When he was not offering questionable medical advice or promoting pseudo-prescriptive remedies to resolve his listeners' ailments, he was hawking a wide array of other goods over the station, such as candy, insurance policies, and electric clocks. One could even procure an "autographed" picture of Jesus at the Last Supper and a wind-up doll of John the Baptist that walked around until its head fell off. Brinkley would sell almost anything to inspire his listeners to send money that further padded the fabulous wealth already generated by his medical practice and radio antics.[6]

For G.W. Coffman and Harvey Averill, Brinkley's lack of taste, quackery, and unsavory business practices were too much. Coffman, the Kansan, insisted that it was time to "forever eliminate this menace, and interference." "You tune in to Brinkley's station, and many times you hear Brinkley himself dishing out medical propaganda," he railed. Did Brinkley's broadcasts, Coffman asked rhetorically, offer anything of substance? "Nothing that amounts to anything, to anyone with brains," he quipped. The Oklahoma newspaperman Averill echoed those complaints. "Frankly, I should much prefer to listen to some good programs than to have to hear of his 'goat gland' treatment or of being told of the operations of some of the Mexican lotteries that are flooding the air with appeals for American dollars."[7]

By the early 1930s, Senator Clarence Dill had emerged as an outspoken critic of the federal government's inability to protect the American airwaves from Brinkley's transgressions. The Democratic chairman of the Commerce Committee, to whom Harvey Averill and G.W. Coffman directed their protests, received increasing numbers of complaints about Brinkley from listeners across the country. Senator Dill, in turn, brought Coffman and Averill's protests to the State Department's attention. Dill identified Coffman and Averill's letters as representative of those he received, and he forwarded them to Under Secretary of State William Castle.[8]

The broad parameters of Brinkley's story that situated him at the heart of the resulting US-Mexican diplomatic dispute has been told many times before. In 1917, the North Carolina native with a dubiously earned medical degree and a checkered past moved to the small central Kansas town of Milford, about 150 miles west of Kansas City. By 1923, with his practice thriving and bolstered by the sales of his pseudo-pharmaceuticals, Brinkley saved enough money to invest in the equipment necessary to establish his own radio station. At a time early in broadcasting's history, when one could often obtain a license just for the asking, Brinkley had little trouble getting KFKB ("Kansas First, Kansas Best," or alternatively "Kansas Farmers Know Best") on the air. He used the station to attract patients to his Milford practice by promoting his signature "goat gland" operation, other quack treatments for a variety of ailments, and his line of pseudo-pharmaceuticals. More money came in from the advertisements he sold. By the end of the 1920s, in the context of increasingly strict regulation of radio and medical licensing, Brinkley's efforts attracted the attention of both the Federal Radio Commission and the American Medical Association. In fact, Brinkley's medical procedures were far from harmless quackery. By 1930, forty-two patients of all ages and backgrounds had died at his clinic in the preceding years. At least six of those deaths were a direct result of an unsuccessful goat gland operation, with many others from complications following that or another procedure. By 1930, Brinkley's antics cost him both his radio and Kansas medical licenses. That outcome motivated him to run for governor as a political independent that same year in a quest to secure the authority necessary to counter his critics.[9]

Brinkley's pivot to Mexico came in the aftermath of his failed 1930 campaign for the Kansas statehouse. He struck a deal to build and operate the powerful station XER in the border city of Villa Acuña. The station's signal easily reached Kansas, as well as many surrounding states. Brinkley capitalized on that transmission range to promote a second 1932 run at the governorship. After losing his medical license in 1930, he initially kept his Milford clinic open and hired credentialed doctors to do his procedures. Brinkley delivered his XER programs from his original Milford, Kansas, radio studio by renting a telephone line that relayed his broadcasts to Mexico for transmission. After his 1932 election loss and spending some $10,000 a month to rent telephone lines for the broadcasts—the equivalent of some $215,000 to $230,000 monthly in 2023—he decided it was time to leave the "dead and so badly isolated" town of Milford for Texas. He capitalized on Texas's laxer licensing standards to secure a medical license in 1933, relocated his clinic and radio studios to Del Rio (the Texas city that sat just opposite Villa Acuña), and survived an attempt by the Texas Medical Board to revoke his new medical license. Securely situated in his Del Rio studio, Brinkley used a much shorter and less expensive telephone link to relay his programs to Villa Acuña. In this way, XER continued to blanket the US airwaves with programs of questionable taste and blatant Brinkley self-promotions. To Brinkley,

Figure 3. Station XER. Courtesy of the Library of Congress Prints and Photographs Division.

it mattered little that he was targeting an American audience with broadcasts transmitted from a foreign country. "Radio waves pay no attention to lines on a map," he confidently asserted. But his professed "Sunshine Station Between the Nations" certainly drew more attention to them.[10]

From the American vantage point, it was not initially evident or obvious that cross-border broadcasting between the United States and Mexico would generate such apparent hostility. A year before Brinkley began broadcasting over XER, Maurice Altaffer, the American consul in the Mexican border city of Nogales, celebrated the plans to build a 1,000-watt station there. Nogales sat adjacent to the Arizona border some seventy miles south of Tucson. "Due to the fact that Nogales is directly on the international boundary," the consul remarked, "it is thought the location of the station will be an unusually good one for reception both in the United States and Mexico." In addition to reaching the west coast of Mexico, Altaffer predicted that the new station would be heard throughout much of the American Southwest, as well. "Such a station should be popular," Altaffer approvingly noted, "and should do much to promote mutual understanding among the people of the two areas."[11]

Brinkley's radio exploits soon exposed Altaffer's naiveté. That the banished American broadcaster returned to the airwaves with an official Mexican blessing added to the outrage. US Ambassador to Mexico J. Reuben Clark implored Augustin Flores, Mexico's acting radio chief in the Ministry of Communications and Public Works, to block Brinkley from the air. Brinkley, the ambassador insisted, was a dual threat to public health and American broadcasting stations. In Clark's estimation, for Mexico to allow its territory to be used by Brinkley to promote his dangerous ideas about medicine to a vast American audience and in the process debilitate legitimately licensed US radio stations was an abdication of "the requirements of good neighborliness."[12]

Clark's pleading speaks to the fundamental vulnerability of the US position. Clark could do little more than beseech Mexico to mind its manners because the United States had no legal right to dictate how Mexico licensed radio stations in its own territory. The United States had not signed any treaties with Mexico regarding the allocation and use of wavelengths. In reality, Clark was sympathetic to Mexico's position. The United States had unilaterally claimed the lion's share of broadcast frequencies in the Western Hemisphere. The ambassador presumed that the ultimate outcome of the conflict would be a conference in which the United States conceded frequencies to Mexico. He figured such a request was wholly reasonable, since "we seem so well spread over the dial" to the point that Mexico had little room to add stations of its own. "I think she might be willing, if properly approached," Clark posited, "to work out a scheme that would interfere as little as possible with us." While Clark was not wrong about the eventual outcome, he certainly was not anticipating that it would take nearly a decade to reach that precise resolution.[13]

In the moment, however, Clark's advice was not heeded. Instead, Vice President Charles Curtis intervened on behalf of Brinkley. Curtis was a fellow Kansan and Brinkley's friend. Prior to his election as vice president in 1928, Curtis was a Kansas senator who rose to Majority Leader in 1924. Curtis's rising political fortunes owed much to the broadcasts he had made over Brinkley's KFKB. In the autumn of 1931, with XER poised to begin operations, Brinkley complained to his influential friend that he was "being hounded in foreign countries" by US authorities bent on "persecuting" him. The State Department, which lacked legal recourse to move against Brinkley, stood down in the aftermath of Curtis's intervention. The path was clear for "Brinkley's grudge station," as one disgruntled but fundamentally correct listener put it, to begin wreaking havoc.[14]

That Mexico allowed the American Brinkley to broadcast in Mexico was the culmination of a century-long story of festering US-Mexican antagonisms. The animosities included the overrunning of then Mexican-controlled Texas by American settlers in the first decades of the nineteenth century, followed by a mid-nineteenth-century war that saw the United States seize half of Mexico's territory, encompassing Texas and the present-day American Southwest, as its prize for victory. The years that followed saw a continuing succession of unwelcome interventions and occupations. These actions unfolded in tandem with an increasingly suffocating US presence over so many aspects of Mexico's overall political and economic development.[15]

By the 1920s, the United States' long-standing proficiency at gaining the upper hand over Mexico seemed to be repeating itself with radio. As broadcasting boomed during that decade, the United States and Canada informally agreed to divvy the hemisphere's frequencies among themselves to minimize cross-border interference between them. This understanding included the designation of 50,000-watt exclusively assigned "clear channel" frequencies that became a fundamental part of the US regulatory structure by the end of the 1920s. The United States claimed forty such channels and Canada took six. That decision greatly limited the remaining number of freely available radio frequencies Mexico could use to develop its own broadcasting system. Into the early 1930s, the United States flatly refused Mexico's requests to clear some space on the spectrum, and in 1932 the Americans and the Canadians turned their informal understanding into a formal agreement without Mexico.[16]

By 1931, Brinkley was a beneficiary of the animosities and hostilities that had defined the previous century of US-Mexican relations. Rather than accept American dictates over hemispheric radio, Mexico saw in Brinkley a potential tool to achieve a better outcome for itself. A Mexican-backed Brinkley could force the United States to the negotiating table on Mexico's terms. And it was not just Brinkley. In the early 1930s, Mexico invited other quacks and rogues whose radio antics had cost them their American station licenses. Beneficiaries included Iowa's Norman Baker, who like Brinkley was a quack doctor and failed gubernatorial candidate whose promotion of an ineffectual and dangerous "cure" for cancer saw him lose his medical and radio licenses and forced from his state. The owners of the Texas-based Crazy Water Crystals Company were also among the radio rogues welcomed into Mexico. This company had used its radio broadcasts to claim its spring water had the ability to cure a range of ills and illnesses as wide as Brinkley's goat-gland surgery. Crazy Water's owners did not wait for the inevitable push off the US airwaves but instead preemptively sought and received an invitation from Mexico. Mexico's strategy was to allow discredited American broadcasters to operate stations on the Mexican side of the US-Mexican border at such high levels of power (two to three times higher than the highest-powered US-based stations) that they were guaranteed to drown out legitimately licensed US stations across a wide swath of states with their unconventional and controversial programming. In that way, Mexican officials hoped, US officials would have no choice but to concede to demands that Mexico receive priority access to frequencies of its own.[17]

The decision to shortchange Mexico on frequency allocation reflected official convictions that radio broadcasting had no realistic prospects to develop in Mexico. Walter Boyle, the US consul in San Luis Potosí, dismissed the potential market in his city and outlying areas on the basis that "primitive" "'peon' and peasant classes" living in the area made for unlikely radio users. Heavily populated Guadalajara and the surrounding region was considered a somewhat better bet for radio's growth, the US consul in that city argued, but only because larger numbers of foreigners and a "better class of Mexicans" lived there than in other parts of Mexico. Even so, poor atmospheric conditions and an economic depression still worked against radio's development. Doubtful of radio's prospects in Mexico, US officials rebuffed Mexican efforts in the mid-1920s to negotiate a reallocation of frequencies that included more for Mexico.[18]

Contrary to such pessimistic predictions, Mexican broadcasting did gain traction during the latter part of the 1920s. In fact, the Mexican state, a product of the Mexican Revolution that erupted in 1910, pursued the development of all forms of radio communications during the 1910s. It saw in radio a means of strengthening its authority over the entirety of a country plagued by regional conflict and enduring revolutionary violence. By 1930 some thirty privately owned commercial broadcasting stations and ten government broadcasting stations blanketed the country. The estimated 100,000-plus receivers owned by individual Mexicans at the end of the 1920s might seem paltry, especially when compared to the more than 62 percent of American households that owned at least one of the approximately eighteen million receivers in the United States by 1933. But the lesser number of Mexican sets presents a misleading picture of that rapidly growing broadcasting market. By 1930, Mexico established itself as the second largest importer of American manufactured radio sets behind Canada, while also seeing its own nascent radio manufacturing industry take root. Beyond individual receiver ownership, the Mexican government and other private Mexican interests distributed receivers to schools, cantinas, and workers' centers across the country, which in turn encouraged radio listening in larger groups. Even the American broadcaster NBC established ties with major Mexican broadcasters, while American advertisers like Coca-Cola and Gillette bought time over a variety of popular stations.[19]

From Mexico's perspective, however, that growth should have been even greater. As Mexico saw it, development of radio in that country suffered because the United States, with Canada's complicity, declared so many of the hemisphere's frequencies off limits to any but American broadcasters. With the growth of its own radio market stunted by persistent interference from the United States, Mexico wanted the United States to cede twelve high-powered clear channel frequencies for exclusive Mexican use. Of the twelve clear channels Mexico demanded as part of any radio agreement with the United States, it insisted six be close to the border. The Americans refused. However, in the absence of any radio agreement with the United States, Mexico was free to act as it wished. Mexico then licensed the so-called border blasters like Brinkley to disrupt US domestic broadcasting. When, during the first year of Brinkley's broadcasts from Mexico, the United States remained adamant in its refusal to yield any ground, Mexico responded by approving XER to broadcast at an astounding 500,000 watts of power starting in August 1932. The increase made XER ten times more powerful than the maximum 50,000-watt station in the United States. That move, Mexico figured, should get the Americans' attention.[20]

It did. That Brinkley was able to remain on the air for nearly a decade is testament to a fortuitous combination of a divided Mexican state and American diplomatic incompetence. Mexican official and public opinion was, in fact, split about permitting Brinkley to operate from Mexico. After nearly a century of overbearing and outright imperialist military, political, and economic interventions, including US domination of Mexico's massive oil industry, Mexican nationalism and anti-Americanism had, in fact, hit a high point by the early 1930s. Would the chaos Brinkley wrought on the American airwaves finally compel the imperious United States to yield to Mexican demands? Or was allowing Brinkley on Mexican airwaves an affront to the rising Mexican nationalism and anti-Americanism of the 1930s? Mexico's most popular daily newspaper, *El Universal*, which had a predominantly middle-class readership, advocated using Brinkley, admittedly a "quack doctor," to fight American "imperialism of the air," even if he did broadcast "improper propaganda" that was "prejudicial to public health." However, according to *La Prensa*, the country's fourth most read newspaper, a position it built upon its devoted base of working-class readers, the situation was intolerable. *La Prensa* lauded the "protesting patriots" who refused to tolerate the farce of claiming that XER was "a Mexican broadcasting station operating in Mexican territory under Mexican laws," all while broadcasting in the "Yankee tongue."[21]

Those contradictory sentiments coexisted at the official level. Prior to securing a license to broadcast in Villa Acuña, Brinkley had been denied such permission from the local authorities in Monterrey. In line with *La Prensa*'s viewpoint, these officials could not countenance the disreputable Yankee broadcaster setting up shop in their locale. Brinkley eventually found willing partners in Villa Acuña, a town located just south of the Rio Grande in the Mexican state of Coahuila. Authorities there gambled that Brinkley could generate economic growth for the city. They pressured the Mexican Communications Ministry to grant Brinkley a license, which it did in 1931. But no sooner did XER go on the air than Mexico began strengthening its radio laws in a way that undermined

Brinkley's position. The revised regulations prohibited foreign ownership of stations, broadcasts from Mexican-based stations that originated from outside Mexico's borders, and the broadcasting of medical advertisements in English, all of which applied to Brinkley. Bolstered by the new regulations, Mexico's public health authorities moved against Brinkley in 1933, hitting XER with steep fines and subsequently moving to have him banished from the airwaves for broadcasting medical advertisements that violated Mexico's radio and public health laws.[22]

By 1934, Brinkley had also alienated Mexico's Ministry of Communications, as well as the Coahuila state government. Both charged that Brinkley was violating the Mexican laws that prohibited him from broadcasting medical advertisements or practicing medicine in Mexico. Newly elected President Lázaro Cárdenas attempted to lend support to the anti-Brinkley effort shortly after taking office in 1934. He authorized the seizure of XER in response to its unpaid fines. Cárdenas, however, was undercut by Mexico's Supreme Court, which ruled that Brinkley was permitted to continue broadcasting at his newly acquired XERA outlet. A Mexican ownership group fronted the new station, which went on the air soon after Mexican authorities seized XER. Brinkley exercised control from behind the scenes. Even the fiercely anti-American Cárdenas, perhaps best remembered for his successful expropriation of American oil and agricultural holdings in the face of fierce US opposition just a few years later, could not prevail over Brinkley.[23]

US officials were no more effective. As powerful lawmakers like Clarence Dill demanded from the Senate floor that Brinkley be stifled, American diplomatic ineptitude further contributed to Brinkley's longevity on the air. Despite sharing with some of the top Mexican leadership the goal of ousting Brinkley, US diplomats could not get out of their own way to exploit that common ground. During the closing months of Herbert Hoover's presidential administration, the State Department chose to pretend Brinkley's escapades were of little consequence to American radio. Viewing the Mexicans as unsophisticated hagglers, US diplomats feared that if they acknowledged the extent of Brinkley's disruptive presence, Mexico would, in the words of Under Secretary of State William Castle, "undoubtedly try to strike a harder bargain than would otherwise be the case." NBC vice president C.W. Horn concurred, insisting the United States would be best served by not admitting to "any anxiety or apprehension" over Brinkley's disruptions.[24]

Franklin Roosevelt's administration, which took office in 1933, stayed the course. Its approach belied the new president's professed commitment to pursuing a "Good Neighbor" approach to US diplomatic relations with Latin America. No such pivot was evidenced in radio policy toward Mexico. That unyielding approach to negotiations predictably contributed to the collapse of the 1933 North and Central American Radio Conference hosted by Mexico. The purpose of that conference was to draft a treaty establishing a regional framework for radio frequency allocation. The treaty, US officials hoped, could have set the parameters for Brinkley's removal. The collapse came after the American delegation incessantly badgered Mexico over its inability to enforce its own radio laws that, the Americans claimed, justified revoking Brinkley's license. The delegation also remained steadfast in its refusal to entertain the possibility of granting Mexico any rights to the clear channel frequencies that the United States already claimed. Such tactics offered little motivation for Mexico to move more vigorously against Brinkley. Brinkley, in turn, was able to stay on the air as the United States and Mexico remained at loggerheads over clear channel frequency allocation. This longevity on the air mobilized many listeners to protest Brinkley and demand the American airwaves be protected from his purported un-American intrusions.[25]

Blasting the Border

Brinkley's XER and its successor station XERA could be heard with ease throughout much of the American Midwest. It could even reach listeners as far as California and New England, albeit with less consistency. The stations offered a programming potpourri of astrologers, fortune tellers, numerologists, "hillbilly" music, and Mexican ballads that surrounded his colorful "medical" talks. The programming was not universally hated. Brinkley's appreciative listeners, mostly in the Midwest, were often representative of those listeners who felt poorly served by the emergent American system of commercialized network radio. They sent volumes of letters to the station praising the broadcasts in general, as well as specific performers. In a story about the controversy Brinkley helped inspire, *Time* magazine noted that thousands of listeners still tuned in to XER daily.[26]

Brinkley's detractors were many as well. Complaints about interference from XER came from nearly every state. Angry letters flooded individual stations, the FRC and its successor agency the FCC, politicians, and the State Department itself. Due to a combination of frequency changes mandated by the Mexican state and the station's technical shortcomings that caused it to deviate from its various assigned frequencies, Brinkley's broadcasts wreaked havoc at several spots between 600kc and 850kc on an AM dial (which ran from 550kc to 1700kc). Popular clear channel stations that enjoyed transmission ranges far beyond their respective home cities, such as Chicago's WGN, Atlanta's WSB, Denver's KOA, Cincinnati's WLW, and New York's WOR, experienced particularly devastating interference. Those clear channel stations, a central pillar of the American system of broadcasting, exposed listeners to programs and significant events that originated from far beyond their particular localities. Many listeners enthusiastically embraced radio's ethereal bridge that connected their locality to distant stations and to the network programs broadcast in markets throughout the country. Part of the thrill of listening to these stations entailed the listeners' recognition that incalculable numbers of other anonymous Americans spread out across the country were listening to the same programming at the same time. In that way, such stations could help foster a sense of national identity.[27]

These protest letters underscored the rogue station's challenge to the national sensibilities that broadcasting had helped cultivate among such listeners. R.C. Keagy, a minister from South Dakota, complained that XER drowned out two or three Chicago stations, as well as other favorites from Cincinnati, New York, and Lincoln, Nebraska. "Its unusual power," Keagy complained in 1931, "smothers a dozen stations to which we are delightfully accustomed." A listener from Hollywood, California, complained in 1931 about XER's deleterious effects on Atlanta's WSB, Chicago's WGN, Cincinnati's WLW, and New York's WOR. The writer pleaded with the Federal Radio Commission to fix the problem so that radio fans "can have a little pleasure fishing for our US stations" without having to endure the interference. A listener in Sidney, Montana, could not believe that Brinkley, a "faker" and "no good Dr" who "just ruins WGN," was even allowed on the air. A listener in St. Louis was "thoroughly disgusted" by the station and the interference it wrought, but his complaint was brief. He correctly surmised that the thousands of other protests undoubtedly pouring in offered more specifics. In 1936, the Lions Club of Newcastle, Wyoming, was livid that XER drowned out Denver's clear channel KOA and several other outlets in its locality. The group complained that the unyielding interference was "a nuisance and is depriving many American families of wholesome entertainment."[28]

The language of these protest letters placed particular focus on the presumably inviolable southern border that separated the United States and Mexico. The winding Rio Grande often symbolized that border. Moreover, these writers penned their objections at a historical moment of skyrocketing anti-Mexican sentiment. Depression-era fears of foreigners taking American jobs singled out unwelcome and purportedly inferior Mexicans as a primary culprit. Amidst this "frenzy of anti-Mexican hysteria," as one study put it, Mexican immigrants and their American-born children faced deportations and forced repatriations to Mexico. At the precise time Americans fought to shore up the notion of an inviolable border that kept out unwanted Mexicans, Mexico-based XER underscored its porousness. In his complaint about the interference generated by XER, Arkansas doctor Edwin Swindler could not believe that Brinkley had been allowed to move south and "is now broadcasting from the Mexican side of the Rio Grande." T.E. Taylor of the Miller-Taylor Shoe Company in Columbus, Georgia, also referenced the Rio Grande border in his complaint about interference afflicting his local station, Atlanta's WSB. "There is a new station, XER in Mexico across the river from Del Rio, Texas," Taylor lamented, "that drowns out Atlanta badly." W.F. Jenkins, a sitting justice on Georgia's state court of appeals, lent an air of authority to his protest by writing on the court's official stationery. "The Mexican station, as I understand it," Justice Jenkins explained, "is some American medicine concern that has been denied a license in this country and stepped over the line." The enraged owner of KMMJ in Clay Center, Nebraska, struck a comparable tone in expressing his disbelief that Brinkley could simply set up operations "just across the line and violate American laws." The listener further underscored his own status as a US citizen in good standing to insist that Brinkley relinquish his citizenship if he refused to comply.[29]

The picture presented of XER illicitly crossing the supposedly inviolable US-Mexican border into sovereign American territory extended beyond impassioned listener protests. A December 1931 *New York Times* article touted a "fight" on the "Mexican Border" provoked by Brinkley's illicit station located "just across the Rio Grande from Del Rio, Texas." In 1934, *Time* magazine referred to "Brinkley's troublesome XER" that lay "across the Mexican border." Three years later, the *Chicago Daily News* professed its certainty that "everybody in the United States has heard the radio stations that cluster on the border opposite Del Rio, Tex." A complaint from Sydney, Montana, explicitly articulated a conviction that the invisible radio spectrum was part of identifiably American national territory when she questioned how Brinkley ever obtained "the right to broadcast and use *our* wavelengths from the Republic of Mexico."[30]

These descriptions of XER's trespasses, with their focus on geography and borders, illuminate the relationship of maps to national identity and nationalism. Maps are the primary means by which a nation's territorial boundaries are visually represented. At the same time, it was the act of listening itself that triggered what media historian Susan Douglas refers to as the "brain's own visual imaging apparatus." When presented with the disembodied sounds coming from the receiver, listeners are compelled to create visualizations of what they believe the people and surrounding context look like. Protest letters about XER's transgressions reflect listeners' geographic visualizations, in turn using language that also compelled the reader to visualize a United States map with a particular focus on the southern border with Mexico. That imaginative ability of both writers and readers depended on possessing knowledge of the nation's geographic expanse, as well as the geography excluded. In this regard, the writers and readers of these protest letters likely had little difficulty putting their imaginations to work in that way. The development of public education in the United States since the earliest days of the republic placed a premium on teaching geography as one of the essential ingredients to foster common notions of citizenship, national unity, and culture. As the uneven quality of early American schooling gave way to educational reform, improved teacher training, and rising public school enrollments during the nineteenth and early twentieth centuries, the numbers of increasingly literate Americans exposed to this geographic knowledge increased exponentially.[31]

For that cross-section of Americans registering their vehement opposition to XER's illicit border crossings, exposure to map-based representations of the American nation continued beyond any formal education they received. Since World War I, newspapers and periodicals increasingly used maps to accompany printed stories; such images engaged geographically rooted notions of the nation and national identity. Indeed, seeking to capitalize on the listening public's desire for national radio during the 1920s, radio publications and advertisements also frequently used maps to encourage listeners to envision broadcasting as encompassing the entire American nation. The cover of the May 1925 issue of *Radio Broadcast*, for example, celebrated the "March of Radio" with a picture of a young boy in front of a map of the continental United States,

Figure 4. Cover image from the May 1925 issue of *Radio Broadcast*.

Figure 5. Phenix radio advertisement from the September 1924 issue of *Radio Broadcast.*

showing his father all the locations across the country from which he had been able to hear a broadcast. A 1924 advertisement from the Phenix Radio Company touted "greater distance on the loudspeaker" by depicting a Phenix receiver against the backdrop of a US map, with each end of the receiver touching the opposite coasts of the United States. A continental map of the United States also accompanied a story published in the September 1927 issue of *Radio News*, which proclaimed the inevitable force of radio to level ethnic differences and spread knowledge of the English language. At the center of the map was a giant radio powered by two batteries representing "Americanism" and "the English Language" that rendered the medium the nation's "New Melting Pot."[32]

This use of maps in radio periodicals to celebrate the national reach of broadcasting captures their effectiveness at communicating the territorial limits of an exclusionary national community. The maps that portrayed the national reach of American radio typically did not depict the Canadian or Mexican territory that sat adjacent to the United States, but instead presented the United States as something of an isolated island. At the same time, by presenting the American nation as a solid undifferentiated geographic block of expansive territory stretching from coast to coast, this type of American national map obscured the tremendous diversity and division that existed (and still exists) within the geographic boundaries of the country. This implicit exaggeration of the harmony among otherwise diverse peoples is another essential component of maps'

Figure 6. Graphic that accompanied Charles M. Adams's article, "Radio and Our Spoken Language" in the September 1927 issue of *Radio News*.

relationship to fostering notions of a common national identity. Map-based representations of national radio make this exaggeration especially evident. Radio historian Alexander Russo demonstrated the point when he juxtaposed the "new melting pot" *Radio News* graphic against an NBC company map indicating very large geographic portions of the United States not effectively served by its national network. Diversity-obscuring maps such as the "new melting pot," which depicted all-encompassing national broadcasting confined within starkly drawn national borders, thereby helped engage and further cultivate understandings of some collective "us" residing within the represented nation. This representation of "us" stood separate from some other collective but unrepresented "them" presumed to exist beyond the sharply illustrated boundaries of the nation.[33]

This "us versus them" dynamic based on geography and nationality infused many of the complaints against Brinkley's XER. One listener from Lyons, Kansas, groaned that XER "obliterated" her entire "neighborhood" of popular radio stations. Referencing the Spanish language programming that Mexican law required Brinkley to provide, her tone echoed that of her contemporaries who opposed all foreign language broadcasting. "As a resident of the good old United States and patriotic citizen, I much prefer to listen to the programs given by home folks," she wrote, rather than encounter them "sandwiched among languages of other countries which few if any of us can translate." With even greater hostility, Burrell Conway of Vicksburg, Mississippi, railed against the

"Mexican intruder" that "completely ruined" WSB's reception. Several writers, including at least one US official, characterized the border-crossing broadcasts as a "menace" that needed to be throttled. An even more hostile reaction came from Charles Trapp, the publisher of a local Topeka, Kansas, newspaper whose ill-will toward Brinkley dated back to the quack doctor's Milford days. "An invasion by radio is crossing the Mexican border," the newsman wrote in September 1932. Trapp expressed his incredulity that the US government refused to act against it. Such action would come "quickly enough . . . if it were airplanes instead of words invading the United States."[34]

Other complainants drew from familiar tropes of Mexican inferiority to express their anger over the radio intrusions. Brinkley's own constant reminders that his station was located south of the border in Mexico alongside a sprinkling of Spanish language music, programming, and announcements no doubt encouraged such listeners to draw on their existing knowledge and prejudices for their protests. H.C.R. Stewart, an agricultural statistician at the University of Missouri, feared that "XER and other Mexican stations contribute generally only low class programs and are the only thorns in a mighty fine arrangement of stations and channels." To Alonzo Williams of Rhode Island's United Electric Railway Company, the fact that Brinkley chose to base his radio station in Mexico was evidence enough that the quack doctor was "carrying on perfectly illicit trade." To Eula Milborn of Tallapoosa, Georgia, "the stuff coming over the air from the Mexican station is offensive, positively vulgar." One self-proclaimed "U.S.A. Radio Fan" insisted that something needed to be done "to prevent this intrusion on our American programs" and "preserve our splendid American programs at the hands of these rebels."[35]

XER, in short, offered a distinct contrast to protesters' notions of legitimate American broadcasting. The act of protesting XER's intrusions on American airwaves reflected an expression of a complainant's investment in the American system of radio. Grievances conveyed the sense of American national identity that underpinned that investment; they illuminated the ways in which the act of listening propelled listeners to visualize their nation, its geography, and the presumed foreign threat manifested by XER. The border blaster may not have paid any heed to lines on a map, as Brinkley noted, but it did compel its listeners to think about how radio signals effortlessly crossed those lines. Listener animosity manifested a sense of entitlement to enjoy the American system of broadcasting without disruption. The fact that this disruption emerged from an invisible trespasser that violated American borders from outside the nation made matters worse.[36]

Meanwhile, the ever-worsening problem of the cross-border radio trespasses and interference became increasingly intolerable as the 1930s progressed.

Brinkley's disruptions continued unabated even when Mexican authorities temporarily shut down XER at mid-decade prior to the reversal of that order by Mexico's Supreme Court. During that interim period, Brinkley's acquisition of Reynosa-based XEAW and additional broadcasts made over XEPN in Piedras Negras enabled him to maintain his presence on the airwaves. Brinkley was also joined on the air by a host of other border blasters in Mexico and disruptive stations from other nations beyond Mexico (such as Cuba). Many such broadcasters sought to bypass the FCC's authority to engage English-speaking American audiences. However, several Spanish language stations formerly based in Southern California and Texas also decamped to the Mexican side of the border as a means of continuing to reach their US-based Spanish-speaking audience following their removal from the American airwaves during the 1930s. With disruptions to American radio worsening and intrusions upon Canadian broadcasting increasing as well, in 1937 US and Canadian negotiators arrived at an international radio conference at Havana, Cuba, finally prepared to give ground to Mexico in exchange for a resolution of the border blaster problem.[37]

The conference produced the North American Regional Broadcasting Agreement (NARBA). This first hemispheric-wide radio agreement was signed by the United States, Mexico, Canada, Cuba, Haiti, and the Dominican Republic. Like their American and Canadian counterparts, Mexican diplomats arrived in Havana willing to scale back some of their earlier demands. In the context of the impressive growth of Mexican broadcasting during the 1930s despite the obstacles it faced—more than eighty stations were on the air by 1937—Mexico calculated that a prolongation of the interference-plagued status quo was now the bigger hinderance to continued growth, more so than the need for the twelve clear channel frequencies it demanded at the outset of the 1930s. The resulting agreement provided Mexico and Canada with six such frequencies, while leaving the United States with priority over thirty-two frequencies.[38]

The good feelings did not last. Both the American and Mexican negotiators arrived home with concerns that too much had been conceded. For the United States, the main worry was that NARBA did not go far enough to force Mexico to remove offending border blasters from the air. From the Mexican side, it was anxieties that its stations remained insufficiently protected from US-based interference. The Mexican concerns stalled ratification in that country. However, in the haste to bring some semblance of order to hemispheric radio based on the accord, the US Senate voted in favor of ratification in June 1938. Only afterward did the United States ask Mexico for further adjustments that addressed its concerns. The advantage again shifted to Mexico, which responded by predicating its ratification on the US willingness to sign a "gentlemen's agreement" that addressed Mexico's remaining anxieties. The efforts to strike such a deal were further complicated by the broader souring of US-Mexican relations, punctuated by Mexican president Lázaro Cárdenas's March 1938 nationalization of American oil and agricultural holdings in Mexico over fierce but ultimately ineffectual US protests. Consequently, when American and Mexican negotiators finally struck an ancillary radio agreement in April 1939 to address their respective concerns, the supplementary agreement still failed to secure the necessary votes for approval in Mexico's National Assembly. The process was further slowed in the face of fierce lobbying by American-operated border broadcasters who hoped to see Mexico scuttle the treaty entirely. In the interim, Mexico allowed Brinkley to remain on the air. "Just a few minutes ago I listened to the most disgusting harangue delivered via the air by that self stiled [*sic*] wizard of medical and surgical skill, Dr. John R. Brinkley of Del Rio, Texas," one listener complained in a letter sent directly to President Roosevelt in January 1938.[39]

By the late 1930s and early 1940s, the perception of threats to American security posed by Brinkley and his border-crossing broadcasts became even more animated and nationalistic. With wars by then raging in Europe and Asia, the United States mobilized its military and defense industries in response to potential threats posed by the belligerents. Brinkley, however, used his broadcasting perch in Mexico to amplify his loud and obnoxious oppositional voice to US defense policy and conscription. Brinkley also added a strong dose of anti-Semitism to his tirades. In a climate of elevated patriotism and increasing intolerance of dissent, one did not need to know that Brinkley's broadcasts originated from Mexico to deem him unpatriotic and a potential security threat. One such listener from Manhattan, New York, seemingly unaware that the offensive broadcasts came from Mexico, nonetheless raged that his talks were "unpatriotic and against the fundamentals of our constitution." For those who did realize Brinkley's harangues came from Mexico, the foreign origins of this threat increased the urgency. Toward that end, the rogue radioman captured the attention of prominent and controversial Unitarian minister Leon Milton Birkhead. He was the founder of the internationalist-minded "Friends of Democracy" organization, which monitored and drew attention to purported anti-democratic propaganda spread within the United States. Birkhead singled out Brinkley's position "on the Mexican border" to impress upon the FCC the urgency of the situation. "Why spend millions of dollars in military preparedness" only to allow Brinkley "to weaken the morale of our people and disrupt national unity?" Birkhead posited. "Decisive and heroic action should be taken," he concluded. "As Americans we have a right to expect such action." At the very least, many other listeners counseled, the United States needed to finalize and implement the long-overdue NARBA agreement with Mexico to remove Brinkley from the air once and for all. "Surely something can be done to get the necessary cooperation from Mexico," the principal at Arcadia High School in Louisiana queried, so that the FCC could silence Brinkley's "almost treasonable talks concerning the foreign policy of this country."[40]

American diplomats begrudgingly arrived at the same conclusion. After nearly a decade of stalemate, the deteriorating international situation after 1939 ultimately drove the Americans to accommodate Mexico's remaining radio demands. Germany's belligerent challenges to the international order that culminated with its September 1939 invasion of Poland, sparking the outbreak of the Second World War in Europe, had reverberations in Mexico. Amidst the sharp deterioration of US-German relations alongside that march into war, Germany also made a concerted push to increase its influence in Mexico at the United States' expense. The growth of German imports into Mexico, German efforts to influence Mexican press coverage in a way that was favorable to Germany, and German intelligence initiatives in Mexico increasingly alarmed American policymakers. To counter the German influence, the United States sought improved US-Mexican military cooperation, the rights to use Mexican naval bases and Mexican airspace, and joint efforts to drive German interests out of Mexico. However, prospects for securing such objectives required concessions to Mexican demands and positions on other outstanding issues, including giving Mexico additional protections from US-based radio interference. Amidst this emergent cooperative dynamic, the United States agreed to a NARBA addendum that provided Mexico with assurances against interference from the US-based stations in exchange for the explicit promise to remove all offending border blasters from the air. For the United States, Mexico's willingness to resist German influence made the price more than worth it.[41]

Brinkley's fall from grace was swift and dramatic. NARBA's long awaited and mutually agreed upon enactment in 1941 deprived Brinkley of his safe haven south of the Rio Grande. As promised, Mexico pulled Brinkley and other now-outlawed border stations from the air, while the US Internal Revenue Service went after Brinkley for tax fraud. In early 1941, the US Postal Service also charged Brinkley with mail fraud in response to the various stunts he used to convince his listeners to send him money. Plagued by the cost of his defense and lawsuits filed against him, the once-wealthy Brinkley declared bankruptcy in 1941. "Now the patient is dead," Brinkley wrote to his wife in July 1941 in reference to XER. "The Sunshine Station Between Nations is gone." Brinkley was not far behind. The loss of his radio station was followed by the rapid failure of his health. "My doctor says I am living on borrowed time," a now bedridden Brinkley wrote in January 1942 after suffering through two massive heart attacks and a leg amputation that never healed properly the previous year. By May 1942, the now-penniless Brinkley was dead, felled by one more heart attack.

His mail fraud case never made it to trial. The heyday of the renegade border blasters had passed, as did the man who more than anyone came to symbolize this controversial chapter in American broadcasting history.[42]

* * *

From 1931 through 1941, Dr. John Brinkley successfully (and literally) stood outside the American system of broadcasting. He owed his perch in Mexico and his longevity in occupying it to the decade-long US-Mexican radio stalemate. US diplomatic bumbling throughout the 1930s helped protect Brinkley from what was, at times, a mutual Mexican and American desire to remove him from the airwaves. Left to his own devices as the stalemate endured, Brinkley reached an audience of millions of Americans, some of whom embraced him and others who despised him. His broadcasts, originating as they did from Mexican territory, excited the imagination of his listeners, who reacted not only to the content of his programs but also to the idea of invisible radio waves crossing international borders at will. For a decade, those broadcasts engaged many listeners' ideas about the American nation, sovereignty, and identity, which in turn fueled much of the opposition to Brinkley's deliberately provocative programming.

That many of his outraged critics came from small rural towns located in the American Midwest, West, and South further complicates prevailing understandings of Brinkley's place in American radio history. In the standard telling, XER and Brinkley's subsequent stations built a loyal and enthusiastic audience from such locales in the "grand tradition of the farmer station," historian Clifford Doerksen writes in his study of rogue radio broadcasters. Those "farm stations," however, had been among those regulated off the air as American radio commercialized by the end of the 1920s; farmers in rural areas that lost such stations were among those most vocal in registering their objections to this imposition of national radio standards that did not appear to serve their interests or tastes. It was Brinkley's return to the airwaves in the 1930s over XER that helped fill the void. His stations' unconventional programming, Gene Fowler and Bill Crawford write in their study of border radio, "proved to be the most popular among the rural listeners in the heart of America who formed the core of Dr. Brinkley's listening audience." However, the voluminous complaints preserved in FCC and State Department files also point to significant opposition to Brinkley in many small towns located hundreds of miles away from any major city. They arrived from such places as Sydney, Montana (population 2,010) and Redfield, South Dakota (population 2,664). Source limitations unfortunately prevent a more fine-tuned assessment of how or why a diverse array of rural listeners had landed on opposite sides of the Brinkley divide. For those who chose to protest Brinkley, however, their grievances do effectively demonstrate how the act of listening to Brinkley's border-crossing broadcasts can engage internalized notions of national identity rooted in the listeners' existing body of geographic knowledge and preexisting convictions about the world around them.[43]

Brinkley's controversial broadcasts shared common ground with the foreign language broadcasts that, as chapter 1 underscored, also elicited hostile reactions from so many American listeners during the 1930s and early 1940s. Both offered a stark contrast to the commercialized radio programming that was by then defined, for good or bad, as the American system of radio. While foreign language broadcasts occupied marginal perches on the frequency spectrum to make way for the purportedly mainstream commercial and network broadcasters, Brinkley simply overpowered some of those most popular stations with devastating interference. In so doing, Brinkley also presented these listeners with radio programs that contrasted with what had become their preferred norm for American radio. In that way Brinkley provoked aggrieved listeners to articulate impassioned defenses of the so-called American system and the American nation that they believed was under foreign attack.

The grievances expressed were often hyperbolic and vastly overstated the supposed threat to the nation. Outraged listeners demonstrated their attachment to and emotional investment in their imagined national community by mobilizing to fight the trespasses of a "menace" that, in reality, did little more than impinge on their ability to enjoy listening to some radio stations. While Brinkley's quack medical treatments posed a genuine threat to his patients' lives, no such threat to life and limb was posed to those who hated his broadcasts. In 1941, American diplomats implicitly acknowledged the superficiality of these imagined radio threats. Eyeing more disruptive security and economic challenges posed by Germany, the ongoing wars unfolding in both Europe and Asia, and the need to improve US-Mexican cooperation to counter German encroachments into Mexico, US diplomats finally acquiesced fully to Mexican radio demands. In the increasingly destabilized global environment of a world at war, a once unacceptable asking price was now a bargain for the American diplomatic establishment. Mexico might not have been as outwardly powerful as the United States. But its deployment of static interference in the form of XER and other border blasters as a diplomatic weapon against the United States leveled the playing field and secured for Mexico a win that was a decade in the making.

PART II

Language, Education, and Identity on the Radio

CHAPTER 3

"To Help the French Speaking People of Louisiana"

Language, Education, and Identity in the French Radio Project at Louisiana State University, 1938–1940

Louisiana State University's French Radio Project was off to a rocky start. Launched in 1938, its purpose was to broadcast French language educational programs to the state's French-speaking Cajun population. The French Radio Project was conceived as a vehicle to engage and improve the lives of the impoverished Cajun listeners in twenty-three predominantly Cajun parishes in Southern Louisiana that collectively had a population of about 600,000 people. The project's potential convinced the philanthropic Rockefeller Foundation to fund the salary and expenses of a fieldworker. The position entailed a responsibility to travel to the parishes, promote the programs, and work directly with participating listeners. LSU alum Louise Olivier got the job. LSU Extension Division director Preston H. Griffith, under whose auspices radio education fell, sang her praises. To the foundation, he touted Louise Olivier's credentials as a "native French woman" (meaning she was of French descent) who was "very clever," "intelligent and enthusiastic." Olivier's "good personality" and "very fine physique" added to those strengths. "I think we are very fortunate in getting this young lady," Griffith offered.[1]

The project's other primary leader, linguist Harley Smith, sharply disagreed with this assessment. Smith, who created and oversaw the project, as well as wrote the language lessons for broadcast, quickly came to despise the fieldworker foisted upon him. "Miss Olivier has obviously never done research work," Smith groused. She had "great difficulty" in understanding "the most elemental directions" and those instructions "given her—in writing—are not carried out or are carried out incompletely." Olivier's incompetence only wrought "confusion and failure," Smith seethed. Worst of all, Smith charged, Olivier was an elitist who believed the "country" people were "ignorant and do not want to be informed." For that reason, he complained, she only traveled to the urban centers of the targeted parishes. Such a disposition, Smith concluded, was contrary to the intellectual foundations of his project. Olivier, he insisted, was ultimately "doing more harm than good." Just weeks into the broadcasts, he refused to work with her any longer.[2]

At the heart of this rift between Smith and Olivier was a philosophical disagreement over language usage. Louisiana had an exceptionally large population of people who spoke French or a Cajun dialect of it. This circumstance reflected Louisiana's unusual history of French settlement dating back to the eighteenth century in the shadow of rising Anglo-American dominance in North America. Dating back to these "Acadian" origins, well into the twentieth century some familiarity with spoken French or Cajun was a fact of life for many of the state's lifelong residents. Contrary to an oft-expressed aversion to foreign language programming expressed outside Louisiana, in this context it was a given that some form of French had to be relied upon in broadcasts. It was the most effective means to engage an American audience that often did not speak English particularly well. The debate instead pivoted around the use of standard French or the Cajun dialect of that language more familiar to the intended audience. Both Smith and Olivier saw the choice as crucial to how effectively radio programs could educate Cajun communities across the state. Smith, who earned his PhD from LSU's Department of Speech in 1936, created lessons and scripts that used Cajun to reach the target audience. It was the pragmatic choice, he argued. Using the familiar everyday dialect spoken by those who listened offered the best opportunity for the educational broadcasts to effectively teach practical skills that improved economic well-being. Olivier disagreed. She joined the project buttressed by beliefs of the Cajun dialect's inferiority. Olivier's 1937 MA thesis on Cajun "variants from standard French" evidenced in one parish was completed under the auspices of LSU's Department of Romance Languages. That academic interest in standard French was one component of her broader discomfort with Cajun identity. From this intellectual foundation, she advocated using radio to promote a purer form of the French language that she deemed central to Louisiana's supposedly rapidly declining French heritage. In stark contrast to Smith's views, Olivier insisted that "proper" French be the sole language used in the instructional broadcasts.[3]

This conflict over language usage hobbled the French Radio Project from its inception. This struggle underscores why an otherwise well-meaning educational broadcasting initiative ultimately had little to show for its effort. For Louise Olivier, who prevailed in the power struggle with Harley Smith just months into the enterprise, the French Radio Project highlighted her determination to celebrate and rejuvenate a socially constructed and historically questionable sense of elite Acadian cultural identity. It was a determination fueled by an oversimplified understanding of Louisiana's history of French settlement. Under Olivier's guidance, the substance of the French Radio Project became a vehicle to advance a vision of Louisiana's French traditions rooted in her preexisting cultural values and worldviews. To Olivier, Cajun represented a bastardization of Louisiana's proud French past. In her view, reestablishing the French language's prominence, therefore, was a key to arresting the Cajun corruption of Louisiana's once laudable Acadian traditions. For Olivier, radio was only a means to an end. Girded by the suspect historical foundation on which her convictions rested, Olivier's outlook and approach were not well suited to accomplishing the overall educational objectives that informed Harley Smith's radio curriculum. The French Radio Project predictably had little chance of achieving its original goals once Olivier undermined its foundational educational premise.

Cajun versus French and LSU's "French Radio Project"

LSU's so-called French Radio Project was linguist Harley Smith's creation. University president James Monroe Smith (no relation to the linguist) and Extension director Preston H. Griffith had approached Smith in 1938 with a nascent idea for developing educational programs that would target the French-speaking sections of Louisiana. With its stated purpose "to help the French speaking people of Louisiana," the venture would provide a mix of engaging informational and entertainment programming. Such an initiative, they reasoned, might better equip this demographic to improve their economic and social standing. Harley Smith was tasked with transforming that inspiration into an executable project.[4]

The linguist jumped at the chance. At the time, Smith was a young and promising academic who recently had completed his LSU dissertation. In developing the French Radio Project, Smith insisted that broadcasts "should start at the level of the listener." Over the course of his subsequent academic career, Smith maintained this counsel in his several books and articles on English language instruction aimed at both native and foreign speakers. His coauthored 1950 text, *Business and Professional Speech*, reiterated this conviction. While not dismissive of the importance of grammatical accuracy, Smith and coauthor Clara Krefting Mawhinney counseled the importance of simplicity and being mindful of the student's existing language abilities. Such an approach, they argued, ensured a student had the time to "develop a few skills well enough that he can put them into immediate use" and "develop at his own speed." Only through cooperation and mutual understanding between teacher and student, Smith and coauthor Ida Lee King Wilbert counseled, could the instructor succeed in helping students to take their rightful place in a dynamic "world mosaic" in which "each small piece must interestingly and carefully be made and must fit into the complete whole." Smith and Wilbert offered that latter advice in their coauthored 1965 text *Everyday English*, which was reprinted well into the 1970s.[5]

Though the broadcasting project he created was French language–based, Smith's pedagogical philosophy for English instruction infused it. For his French Project to achieve its educational goals, Harley Smith understood he had to engage this target audience of French-speaking "country people" on its own terms. Doing so dictated "that the French used on the programs should be, predominantly, the French spoken by the Cajuns of Louisiana." This approach meant using "many idiomatic expressions to give skits a feeling of naturalness." Lessons and lectures delivered in that spoken language, Smith concluded, could provide these typically undereducated peoples with the tools necessary to surmount the array of agricultural, community, and personal problems that often kept the predominantly rural parishes mired in poverty.[6]

That this program was developed at a major university like Louisiana State reflected the nature of educational radio by the late 1930s. Visions of harnessing radio for the purposes of education and uplift were part and parcel of broadcasting's emergence. By 1939, the so-called land grant universities, of which LSU was one, had positioned themselves at the forefront of educational radio. Established through a succession of federal laws passed between 1862 and 1914, land grant colleges and universities were the foundation of publicly supported higher education in the United States. These institutions prioritized making practical education and information accessible to their respective states' residents. These schools pursued that goal not just through traditional campus-based degree programs, but also through so-called extension services that sought to improve the lives of often disadvantaged communities not formally connected to the university. Given broadcasting's ability to reach a large and widely dispersed audience, those priorities help explain why the extension services of so many land grant schools pursued educational radio initiatives. In the 1920s, land grant schools such as Iowa State College of Agriculture and Mechanical Arts, Kansas State Agricultural College, Ohio State University, and Oregon State College had pioneered the development of educational and public service broadcasting to surrounding rural communities through their respective extension services. These college- and university-operated stations offered the types of educational and outreach programs typically not found on the commercial outlets.[7]

These institutions' pioneering efforts notwithstanding, by the early 1930s their position as the pillars of American educational broadcasting was far from assured in the context of ongoing commercialization. In the 1920s, a disorganized cadre of educational broadcasters, including the land grant schools, muddled through their efforts with inadequate funding, irregular schedules, and ineffective strategies poorly suited to attracting and engaging large audiences. It became difficult to claim such efforts effectively served the "public interest" stipulated by the regulatory structure put into place by the late 1920s. Consequently, as the American system of commercial broadcasting secured its dominant position through that regulatory framework, many public service–type broadcasters found themselves squeezed off the airwaves. Licensing decisions forced many of them off the air altogether. Many of the educational broadcasters permitted to remain on the air were nonetheless compelled to reduce their power and switch to less desirable frequencies that were to be shared with multiple other smaller broadcasters.[8]

Faced with a mortal threat to their survival, the once-disorganized collective of educational broadcasters and its advocates worked to define common standards and best practices that satisfied the "public interest" licensing stipulation. As commercialization overwhelmed the airwaves, nine organizations committed to promoting educational radio, heavily represented by land grant universities, combined forces in the National Council for Educational Radio (NCER). They collaborated with sympathetic officials on the Federal Radio Commission, its successor the Federal Communications Commission, and the Department of the Interior's Office of Education (OOE). Sympathetic officials in these agencies genuinely believed the public interest would be served by affirming a place on the spectrum for nonprofit educational radio capable of achieving measurable results. Toward this end, the Federal Communications Commission established the Federal Radio Education Commission (FREC) in 1935, and tapped the OOE's commissioner John Studebaker to lead it. The FREC and the Office of Education, in turn, worked closely with the NCER, especially the NCER's component organization, the National Association for Educational Broadcasters, to develop effective standards and strategies for educational radio.[9]

It all may have been for naught had the underfunded OOE and FREC not cultivated philanthropic support from the deep-pocketed Rockefeller Foundation. With a mission to "promote the well-being of mankind throughout the world," the organization was established in 1913 by Standard Oil's founder and longtime chairman John D. Rockefeller using the immense wealth he accrued from his company's stranglehold over the oil industry. Striving to solve the causes of the multiplying social problems that accompanied industrialization and contributed to poverty, disorder, and instability, the foundation funded globe-spanning projects in medicine, natural sciences, social sciences, and humanities deemed germane to the pursuit of those objectives. By the mid-1930s, the foundation's participation in a series of FCC/OOE/FREC-sponsored conferences impressed upon key officials the relevance of radio education to furthering their philanthropic organization's overarching goals. Rockefeller officials also understood that the prospects for success depended on developing more effective radio education initiatives to secure their place on the otherwise commercialized American radio dial. By the latter part of the 1930s, the Rockefeller Foundation through its Humanities Division was committing substantial financial resources to develop educational radio. This support was reflected in funding provided directly to the FREC as well as to the specific radio projects devised by educational institutions. With much of that money ultimately directed toward land grant universities' radio projects, those institutions established their position at the forefront of public service educational broadcasting.[10]

LSU needed such financial support. The university was behind its peers in the pursuit of educational radio. The French Radio Project, in fact, reflected part of LSU's effort to catch up to its fellow land grant institutions in harnessing radio's educational potential. The 1938 development of the initiative occurred in the context of the university president's decision to combine the institution's disparate radio activities under the singular auspices of its Extension Division. The move was intended to foster a more systematic and sustainable approach to developing educational radio programs for constituencies across the state. The university radio station struck retransmission agreements with other stations to expand their reach. These stations collectively comprised Louisiana's "Southern Broadcasting Network" and ensured its educational radio programs like the French Radio Project could be heard far beyond LSU's home base in Baton Rouge.[11]

As Smith conceived his project, a multilingual fieldworker who could communicate in standard French, its Cajun dialect, and English was critical. That person's activities would connect the targeted communities to educators and the French Radio Project at LSU. Some 80 percent of the 600,000 people in those parishes spoke Cajun. The fieldworker's responsibilities entailed "preparing the French population for listening in on these broadcasts." It was incumbent on the fieldworker to "explain, promote, and interpret the broadcast program" to its various constituencies "as directed by the central office." This mentorship was intended to help listeners acquire the skill sets necessary to take advantage of the information and knowledge the broadcasts offered. The fieldworker was also tasked with communicating assessments to LSU and the foundation documenting how well the programs engaged their targeted audience. These assessments, based on the fieldworker's "actual contact with the people," would give the principal educators and broadcasters the information they needed to develop "the type of program that will assure success."[12]

This approach was meant to address common critiques of educational broadcasting's shortcomings that abounded by the 1930s. These critiques often blamed the lack of results on inadequate preparation resulting in low quality programs, lack of cooperation between educators and broadcasters, inadequate preparation of the target audiences by the broadcasters and educators, and a dearth of systematic evaluations of listener response. That appraisal suffused a 1932 report on radio's educational limitations from the National Advisory Council on Radio in Education (NACRE). This New York City-based group with ties to the commercial broadcasters was a rival to the NCER. In 1935, the University Broadcasting Council (a consortium of midwestern universities led by the University of Chicago that was independent of but viewed favorably by the NCER) struck the same tone in concluding that "educational programs have been poorly designed for radio consumption." Harley Smith sought to surmount these limitations by fostering personal and direct engagement between the broadcasters, educators, the fieldworker, and the targeted audience. Only then, as Smith saw it, could his French Radio Project have a chance to realize popular but often unfulfilled visions of radio's ability break down barriers by spreading knowledge and culture to the benefit of Louisiana's Cajun communities.[13]

This objective was ambitious and promised to be difficult to achieve under any circumstances. Marginalized French-speaking communities in Louisiana traced their roots back nearly a century and a half earlier. The original French Canadians (from which the term "Acadian" is derived) arrived in then Spanish-controlled western Louisiana in 1763. They had been ousted from Canada when Britain wrested that territory from France that same year. Spain, then allied with France, welcomed the settlement of the French Canadians. This infusion of French-speaking peoples, Spanish authorities reasoned, could help economically develop the territory and position it as a buffer against the increasingly strong British Empire in North America. The subsequent arrival of a diverse array of peoples from other cultures and nationalities over the next century-plus introduced new influences and practices into Acadian vernacular and popular culture. These groups included Spanish-speaking officials, free French-speaking Blacks who had migrated from Haiti, African slaves, Native Americans, free "Creoles" of color (peoples of mixed French, Spanish, and African ancestry), and—after the newly independent United States acquired the territory in 1803—other non-English-speaking immigrants. The resulting "Cajun" culture (the term "Cajun" itself originating as a linguistic transmutation of the term "Acadian," first to 'cadian and finally to Cajun) was increasingly distant and in many ways disconnected from its French-Canadian roots. This disconnect included the fact that many speakers of Cajun had no familial or ancestral ties to those original French-Canadian settlers. By the 1930s, Louisiana's Cajuns, among the poorest, least educated, and most marginalized peoples of the region, were largely comprised of subsistence-level small farmers, swamp fishermen, shrimpers, and oystermen. High levels of poverty and illiteracy shaped the development of insular community structures reliant on extended kin groups. Such community structures helped residents navigate their economic hardships and lack of political power, while at the same time fostered distinct forms of language, cuisine, leisure, and Catholicism that signified their distinct identity.[14]

Using broadcasting to reach the more isolated Cajun communities made sense. In the medium's early days, radio quickly established its popularity among the region's largely illiterate lower classes. For example, traditional evening social gatherings focused on food, conversation, singing, and storytelling—*veillées*—proved adaptable to radio and its propensity to encourage group listening. As was the case with newly arrived and impoverished immigrant groups in bigger cities, radio fed the Cajun community's comparable thirst to hear culturally specific programming that reflected its deeply rooted oral traditions. In that way, broadcasting helped strengthen family and community ties. Even though rates of family radio ownership were predictably low among the most impoverished in these communities, group listening in public spaces gained traction, just as it did in poorer communities discovering radio elsewhere in the United States and the world. With these Louisiana communities embracing broadcasting because of its ability to engage and adapt to Cajun cultural practices and ideals, a radio educational plan that sought to reach these people in their own spoken language seemed promising.[15]

With Smith's plan in place, Griffith approached the Rockefeller Foundation in 1938 for financial support to launch the initiative. Since already-salaried faculty developed the programs to be broadcast over the university's existing radio network, LSU could cover most expenses. The additional salary of the fieldworker, however, fell beyond the Extension's budget. The foundation's overall philanthropic objectives and its specific interest in funding educational radio complemented the French Radio Project's objective to alleviate difficulties and tribulations afflicting Louisiana's Cajun communities. The project would do so, Griffith explained, by engaging its target audience "in their own language, on their own level of interest and understanding." More specific to the LSU proposal, the selection of Harley Smith to develop and lead the project was well suited to earn the foundation's confidence. The foundation, having previously funded Smith, was familiar with his work and officials there thought highly of the linguist.[16]

The overall project and its focus on the listener's perspective was particularly appealing to John Marshall, the foundation's assistant director for Humanities. Marshall had become perhaps the foundation's strongest advocate for funding radio education initiatives. Born into a long-standing elite New England family, the Harvard-educated Marshall earned undergraduate and graduate degrees in English, with an emphasis in Medieval Studies. After spending five years in academia, he joined the foundation in 1933. In his assistant director position, Marshall revamped the focus of the Humanities Division and prioritized support for the emerging academic discipline of communications studies. Soon thereafter, his participation in the FCC/OOE/FREC-sponsored conferences of the mid-1930s galvanized his commitment to fund educational broadcasting endeavors. Under his stewardship, the foundation supported an array of educational radio initiatives, including the Rocky Mountain Radio Council and the Chicago-based NCER-allied University Broadcasting Council, to name just two. The latter group's influential critique of educational broadcasting's shortcomings can be found in the foundation's files among Marshall's notes and records. Marshall was also involved with the foundation's 1936 decision to provide funding for the Princeton Radio Research Project. The prestigious Princeton group conducted systematic audience studies. It prioritized research into people's motivations for listening and the psychological value that radio held for its audience. By further describing the role of a fieldworker as one that would overcome the historical shortcomings of educational radio and be mindful of audience engagement, Griffith hit on Marshall's main concerns when it came to evaluating the quality of proposals.[17]

The pitch worked. With Marshall's backing, the foundation agreed to fund the project for two years under the auspices of its General Education Board's Southern Education Project. In explaining its enthusiasm for the project, the General Education Board singled out radio's unique ability to penetrate the undeveloped, inaccessible, waterlogged bayous of the state's Gulf Coast. Radio, the board noted, "affords the best opportunity to reach this group" of listeners with substantive educational instruction. During those two years, the foundation committed to fund the fieldworker's $1,800 annual salary for a total of $3,600 with an additional $2,400 to cover related travel expenses and clerical work. The total $6,000 commitment was the equivalent of about $130,000 in 2023.[18]

Louisiana's French Tradition and Linguistic Purism

Louise Olivier joined the project as its fieldworker in June 1938. The dynamic thirty-three-year-old came highly recommended by her former faculty mentors in LSU's Department of Romance Languages. Less than a year earlier she had completed her Master of Arts degree in French alongside a minor in music. Prior to joining the radio project, she taught French in the public schools and at LSU. Olivier hailed from the small Cajun village of Grand Coteau and was born into an economically established family that prioritized education. She received her early schooling at a Catholic convent near her village, and ultimately finished her secondary education in New Orleans before going on to earn her advanced degrees at LSU. One of nine children, her home was always bustling with people when she was a young girl. Anyone visiting at mealtime, one family friend recalled, was inevitably fed thanks to the family's personal cook, who always prepared enough food "to feed an army," leaving abundant leftovers to be sent home with others. "If they had run a soup kitchen they would have done well." The personable Olivier, known to her friends as Lulu, was recalled by one confidante as "very lively" and "always full of jokes." Her sharp wit, good humor, and infectious laugh made her, in the words of another longtime friend, "the life of any party of which she was a member." She parlayed her gregarious personality and language proficiencies into forging an expansive network of personal and professional contacts that she could tap as the French Radio Project's fieldworker. Griffith had every reason to be excited by the background, credentials, and endorsements Louise Olivier brought to the initiative.[19]

Olivier had, in fact, joined the French Radio Project at a critical historical moment. By the late 1930s, nostalgia for Louisiana's supposedly waning "traditional" French customs was especially strong. Since 1921, when Louisiana's constitution established English as the state's official language in an aggressive effort to expunge French and other non-English languages from educational institutions, spoken French had indeed declined throughout the state. By the late 1930s, more and more children of French descent were speaking English as their first language and, by the mid-1940s, most speakers of Louisiana French had never heard what Louise Olivier deemed to be proper French. Fears that the impingements of the modernizing world had wrought further destruction on tradition amplified growing concerns of cultural decline. The development of transportation infrastructure and new employment opportunities in the state's expanding oil and natural gas industries, for example, increased the exchanges and interactions between English-speaking and French-descended communities. Such accelerating contacts, at least in the eyes of many cultural activists, worked to the detriment of preserving a more traditional French-flavored past.

New technologies intruding into the home compounded angst over cultural loss. The increasing use of refrigerators, for example, seemingly threatened the practice and the cultural significance of communal *boucheries* (the butchering of a pig or cow). Broadcasting, in the eyes of other nostalgia-driven cultural activists, was not so much capitalizing on the traditional form of storytelling known as *veillées*, but rather destroying its significance through debilitating adaptations.[20]

Olivier's experiences growing up with spoken French in her own family intersected with this rising wave of nostalgia for an idealized vision of Louisiana's French past. Despite the relative economic advantages her family enjoyed and their general ability to understand standard French when spoken to them, they struggled to speak it. In Olivier's eyes, it was an unfortunate and embarrassing consequence of the extent to which Cajun influences had corrupted their spoken language. Olivier's mother was an especially poor speaker of French. Other family members recalled their own reticence to speak "bad French" for fear of corrupting the next generation. Olivier's commitment to standard French, not just in the broadcasts but in her life, reflected these formative experiences and became a central part of her personal identity. It motivated her pursuit of a graduate degree in French at LSU. In this regard, Olivier aligned herself with the small cadre of bilingual, educated, urban-minded, and upwardly mobile French Canadian–descended elite that deliberately distanced themselves from any outward appearance of being Cajun.[21]

This upbringing, education, and outlook rendered Olivier disconnected from the poorer Cajun communities the radio project and her related fieldwork intended to target. It was not just a consequence of her advanced education and language preferences. To Olivier, the idea of "Cajun" threatened the superior French Acadian culture that she sought to rejuvenate. Olivier, in fact, obscured her own Cajun background. The Cajun label came with pervasive negative stereotypes and assumptions about the demographic. In this view, it was the stubborn Cajun insistence on clinging to an inferior culture and its practices that was responsible for their marginalization and hardships, which produced the intractable poverty and illiteracy afflicting Cajun communities. The refusal to change course given the pervasive difficulties, in turn, confirmed for many the negative stereotype of Cajuns as cliquish, simple-minded, and lazy. Olivier's many grievances against Cajun culture and practices included her disdain for the supposed southwestern "cowboy" influence on Cajun music. She scorned it as "cow-jun" music, an unacceptable hybrid that she deemed inauthentically Acadian. "One of the peculiar psychological quirks which outside writers will have to learn about South Louisiana," wrote a friend and fellow native of the area, "is that while Miss Olivier will call herself a Cajun and while I can give myself the name, both of us will bristle with ire if anyone else so much tries to call us Cajuns." Underscoring her determination to distance her public persona from the Cajun label, Olivier insisted that a radio program she developed and hosted for LSU separate from Harley Smith's project be called "The Creole Hour."[22]

That word choice captures the inherent subjectivity and malleability of notions of identity and language that informed Olivier's sense of self. "Creole," in fact, long had multiple meanings in Louisiana. To Olivier, it was intended to refer to the white, educated, and upper-class descendants of French settlers with whom she most closely identified. However, other permutations of the term allowed for the inclusion of peoples of Spanish, German, Irish, and African descent, while the dialect of Creole French traced its roots to the seventeenth-century interactions between African slaves and French-speaking Europeans. In that context, the distinction that Olivier drew between Cajun and Creole was not necessarily one of meaningful difference, but rather one that was subjectively and artificially created. The two groups' ancestries overlapped and, according to some scholars, there is effectively no substantive linguistic distinction between Cajun and Creole. Many of the cultural and linguistic practices Olivier castigated as a Cajun bastardization of purportedly superior French ones could not, in fact, be so easily disentangled from what she and others described favorably as Creole influences. In sum, Olivier's use of the term Creole over Cajun in the title of her radio program did not necessarily convey the meaning she intended to all those who heard it.[23]

Though well meaning and sincere, Olivier was, in fact, invested in the lore of French Louisiana's Acadian origins. It was a myth based, at least partly, on the linguistic transmutation that saw the original term "Acadian" morph into "Cajun." The alteration implies an oversimplified Cajun connection to the original Acadian migration. It obscures the verity that multiple demographic groups, including many Louisianans with non-Acadian French origins, contributed to the evolution of Cajun culture and language. Olivier's conception of Cajun as a linear degradation from its superior Acadian predecessor thereby ignores the multitude of influences that shaped the full tapestry of Cajun culture across Louisiana. By not appreciating these underlying complexities, Olivier was not well equipped to achieve the outcomes she or the French Radio Project desired.[24]

That Olivier was inattentive to the historical and practical complexities embedded in the term Creole is further accentuated by her linguistic purism. A linguistic purist seeks to purge a language of perceived undesirable, foreign, or vulgar elements that supposedly coarsen and compromise its quality. It is a value-laden endeavor that presumes the superiority of the supposedly purist form of the language in question over the purported bastardizations and mutations that produced allegedly inferior dialects. More than just a singular fear of language decay, linguistic purism reflects anxieties that the emergence of an inferior form of a supposedly superior standard language also signals cultural inferiority and decline. This commitment to purism tends to be strongest among nonlinguists, and most often originates among a small intellectual elite with an investment in elite culture. The economically secure and advanced degree–holding Olivier fit that description. For Olivier, her embrace of standard French reflected her deep disdain for the supposedly inferior Cajun dialect, which she perceived as degrading Acadian culture and traditions. As such, she approached her work in the French Radio Project as an educational exercise in preserving the French language and, by extension, promoting her vision of a French Acadian culture superior to its inferior Cajun mutation.[25]

Olivier's linguistic purism thus dovetailed with her commitment to a "standard language ideology." She insisted that the many dialectic variations and colloquialisms of Cajun spoken across many different communities could be difficult for all prospective listeners to understand. Using colloquialisms of some Cajun dialects, she warned, might not resonate among those speaking other dialects of Cajun. Though she was not wrong about the varied dialects that fell under the Cajun canopy, she overstated the point to achieve her larger purpose. While visiting various parishes in preparation for the forthcoming broadcasts, she claimed that the people she visited uniformly echoed her sentiments. This judgment was suspect, given that she limited her visits to the urban centers of selected parishes where standard French had greater traction among typically better educated residents. Her convictions rested on a substantial base of misinformation and misrepresentation of Louisiana's supposed French traditions, which informed her highly selective assessment of the wants and needs of the targeted Cajun parishes.[26]

Olivier's skewed understanding of the contexts in which she worked captures how her convictions about a superior Acadian identity and culture had a distortive effect on the French Radio Project. She insisted that her privileged and supposedly unrivaled understanding of Louisiana's French history and traditions infuse the initiative. Ultimately, the cultural authority Olivier wielded as a "native French woman" overwhelmed Smith's vision. Though the initial broadcasts used Cajun, as Smith intended, Olivier immediately began lobbying for a change to standard French. These machinations capture how the application of a standard language ideology to education can underscore and perpetuate a disadvantaged group's systemic lack of access to power and privilege. The impoverished and undereducated target Cajun audience largely did not speak standard French. If Olivier had her way, the target audience would be ill-equipped to understand and capitalize on the knowledge being purveyed through programs supposedly conceived to help improve their disadvantaged positions.[27]

It is hard to imagine a more intellectually ill-fitted pairing than Smith and Olivier. In contrast to Olivier's proud linguistic and cultural purism, Harley Smith embraced the less value-laden approaches and methods more typical of linguists. Linguists tend to steer clear of affixing value judgments that make claims for the superiority or inferiority of different dialects. To the linguist, dialects simply reflect the infinite adaptability of all languages. In this view, dialects cannot be superior or inferior to one another. Rather, every dialect is a highly developed and culturally rooted linguistic tool that meets needs of the community in which it arose and continues to evolve as those community needs do the same. Smith's insistence that Cajun be used in the broadcasts to provide the knowledge that could help the audience improve their circumstances reflected that very sentiment. It was also a sentiment bound to clash with the very different priorities Olivier brought to the project.[28]

Given their obviously irreconcilable views over language usage, it makes little sense that Harley Smith would hire Louise Olivier to serve in such a critical role for an initiative he created. In fact, he did not. University president James Monroe Smith was the final arbiter. He had taken the lead to place LSU's varied radio initiatives under the singular auspices of the Extension Division and create the Southern Broadcasting Network. In this context, it was President Smith in cooperation with Extension director Griffith who hired the linguist Smith to develop the project, and then submitted the first formal request for funding to the Rockefeller Foundation in spring 1938 under his name and signature. Finally, upon receiving confirmation that the foundation approved funding for the French Project, it was President Smith again in conjunction with Griffith who hired Louise Olivier. In articulating his frustrations with Olivier, Harley Smith made clear that he did not choose to hire her and that she had been imposed on him and his project.[29]

Not surprisingly, the bitter conflict over language usage between the two principals erupted almost immediately. The project as conceived by Smith relied on applying two techniques: speeches and dramatizations. Smith, in fact, supported the use of standard French as the preferred language for announcements, special events, and the educational speeches and lectures delivered on practical topics. Cajun, however, was Smith's preferred choice for the dramatizations. Smith conceived of the dramatizations, which comprised the bulk of the programming, as "presenting characters typical of the French speaking areas in typical situations" whose speech would reflect "predominantly, the French spoken by the Cajuns of Louisiana." Olivier not only objected to this premise but lobbied fiercely against the use of any non-standard French in the broadcasts. Toward that end, she did not do as instructed, which was to first establish the listening centers. Smith conceived of these centers as the essential tool through which the fieldworker would engage and mentor local audiences who listened to the broadcasts. She instead spent the autumn and winter of 1938 undermining Smith. She did so by soliciting adverse reactions to the programs from the parishes she visited and then going over Smith's head by reporting the complaints to Griffith.[30]

Harley Smith was livid. As he saw it, Olivier blatantly and deliberately undermined the project's intellectual foundation. Instead, as Griffith himself acknowledged, she primarily focused on soliciting input from "prominent citizens in the French parishes," who like Olivier predictably recoiled at the use of Cajun. To Smith, that disregard for a core component of the project rendered her supposedly evidence-based objections to the use of Cajun meaningless, even malicious. It was this failing that provoked Smith's allegations of Olivier's incompetence and inability to follow directions. On what basis could Olivier know the project was doomed to fail and take such claims to Griffith in the absence of any meaningful effort to execute the fieldworker's assigned responsibilities? Even more infuriating, Olivier's claims that listeners objected to the use of Cajun in the broadcasts baldly contradicted the favorable listener responses that Smith received after the inaugural broadcast. "I am especially glad that the French spoken was not the Parisian and I was able to understand it," one such listener professed. "Here's hoping you have many more such programs to follow." Smith was not so stubborn as to believe he had created an initiative free of any flaws or shortcomings. He did, however, believe that the project as originally conceived "must be given a trial and changes made only after representative groups have heard a number of programs."[31]

No such trial was given. Harley Smith was effectively forced out of the project in January 1939 after only five broadcasts. The professors of French at LSU lined up to support Olivier's position. This push to reframe the initiative around the pursuit of linguistic purity using standard French convinced Griffith to back Olivier. Future scripts, the Extension director ordered, could only use "simple but correct French." In explaining the unexpected change to President Smith and the Rockefeller Foundation, Griffith said it was necessary because of the linguist Smith's refusal to disavow his support for "incorrect type of French that is used among the rank and file of the French speaking people." Minus the value judgment captured by the phrase "incorrect type of French," using the Cajun dialect instead of standard French was the fundamental point of the French Radio Project. With the central premise of his project undermined, Smith had little choice but to resign. He informed the Rockefeller Foundation's John Marshall, Extension director Griffith, and the university president James

Monroe Smith that his position in the project had become untenable. "I do not desire to accept responsibility for fundamental changes," the linguist Smith told President Smith upon quitting the project.[32]

Louise Olivier's Cultural Activism from Radio to Handicrafts

At the start of 1939, after only five broadcasts, the French Radio Project was on the precipice of collapse. The foundation's confidence in Harley Smith to lead the project loomed large in its decision to fund it, and now just months into the initiative the linguist had been forced out. The foundation's Leo M. Favrot, a Louisiana native, reached out to the LSU president to share his concerns over the events that had unfolded. Favrot drew on his Louisiana background to try to make sense of the circumstances that drove Harley Smith to resign. After reading some scripts, he speculated that a slate of characters with colloquial names such as "Tante Alice" and "Noncle Phillippe" inspired hostility from the likes of Olivier. These characters, Favrot postulated, "are made to speak in such a manner that the authorities fear too much emphasis is placed on their provincial manner with possible offense to the French speaking group." That insight, however, did not shake Favrot's faith in Harley Smith's vision for the radio project. Favrot was also sympathetic to the linguist's complaint that development of the scripts and broadcasts had been scattered among too many people without a clear central decision-making authority. Harley Smith created the project and had the necessary training to handle the broadcasting work, Favrot noted, so why had he not received that final authority? "He should have an important place in the direction of the broadcasts," Favrot told President Smith.[33]

In January 1939, two weeks after Harley Smith's resignation, two high-ranking Rockefeller officials arrived at LSU to try to make sense of the mess. Foundation vice president and director for Southern Education Albert Mann and General Education Board associate director Jackson Davis ultimately departed from the university without satisfactory answers. Upon interviewing Harley Smith, they found him in a "highly critical and emotional mood." The linguist laid the blame for the program's struggles squarely on Olivier's "incompetence." From the university's director of broadcasting Ralph Steetle, the visitors learned that, with Smith's resignation, "the bottom dropped out" of the initiative. Steetle affirmed that he still had four recorded programs scheduled for future broadcast. That said, the broadcasting director had little confidence in the last two since they solely used standard French. Echoing Smith's contention that Cajun French was more likely than standard French to be understood by both rural and city listeners, Steetle doubted those last two programs would satisfactorily engage the intended audience. He noted that the sampling of listener responses brought to his attention demonstrated the programs delivered in Cajun French had been, contrary to Olivier's allegations, well received. Amidst these claims, LSU president James Monroe Smith assured the Rockefeller representatives that he had ordered Griffith to investigate the matter more thoroughly and report back to him.[34]

Griffith's response yielded no ground to his critics. The resulting report, issued at the end of January, reiterated Griffith's complete confidence in Olivier. Affirming his support for Olivier's linguistic purism, Griffith flatly contradicted Steetle by repeating Olivier's insistence that listeners objected to Cajun French. He even went so far as to claim that Ralph Steetle had assured him, contrary to what Steetle himself told foundation representatives during their visit only days earlier, that "they would have no difficulty in preparing programs for future broadcast." Griffith then circled back to Leo Favrot in May 1939 to reaffirm his confidence in Louise Olivier as the project concluded its first year. Griffith confirmed that, in his view, the correct decision had been made to support her over Harley Smith. "We have been very fortunate in our Field Worker, Miss Louise Olivier," Griffith assured Favrot. "In addition to being well educated in French and Music, she has a most dynamic personality, untiring energy and a great desire to get the French program over to the French people in the most effective way." Olivier, Griffith declared, had been singularly responsible for the "great success of the project." He noted that Olivier had just completed the foundation's mandated annual report for the project. That document, Griffith promised, would undoubtedly confirm this glowing assessment of the field-worker's efforts.[35]

It did not. In fact, Olivier's own description of her efforts essentially validated Harley Smith's criticisms. Her report showed little interest in the responsibilities ascribed to the fieldworker position. Contrary to her mandate, she only offered a superficial comment on listener engagement. Speaking to the point of radio ownership and habits of group listening, she presented as evidence her personal observation that "almost every other family owns a radio." It was a grossly exaggerated claim easily discredited by the statistics of radio ownership Griffith had previously forwarded to the foundation. She also provided only vague reference to rural listeners without radios who "usually congregate" to listen in groups, without any further comment on how these listeners responded to the radio broadcasts. Olivier devoted only one paragraph in the five-page single-spaced document to discussing program content, still without attention to listener engagement.[36]

Olivier nonetheless professed her supreme confidence that the initiative she shaped in the wake of Harley Smith's resignation was in no need of adjustment. An "enthusiastic spirit" suffused the supposed listening centers, she insisted.

That reaction, however, was not offered in response to the radio programs. In fact, very little time was spent listening to and discussing the fifteen-minute radio broadcasts at the centers. Exercises in cultural preservation dominated the gatherings. Participants, for example, acted out skits and performances rooted in their presumed shared "French heritage." Absent any connection to the radio component she supposedly oversaw, Olivier effused over the enjoyable encounters she had with music, literature, and other cultural expressions during her travels to Cajun parishes across Louisiana. She talked of her efforts to bond with the "old folks" and learn their games, music, and dances, which might be passed on to the younger generation and "encourage a more harmonious family spirit." She told a story of a hoe-carrying elderly farmer who joined one such gathering already underway. He supposedly threw his hoe down upon entering and shouted in French that "it is impossible to destroy the French language!" The old man, as Olivier told it, then proceeded to out-sing and out-speak everyone else in attendance for the duration of the evening. These gatherings of "French people," Olivier proclaimed, never lasted less than two hours and firmly convinced her that these communities "were ready for a renaissance of their original language." One local newspaper reporting on one of these events ascribed to Olivier a determination to "rekindle in the hearts of the French people a patriotic love for the mother language to inspire the present generation with a curious interest in their French ancestral heritage." "The only unfavorable criticism ever made" of the gatherings, Olivier insisted, "is that they were too short."[37]

These activities, however, were wholly disconnected from the project's original and more precisely conceived educational objectives. In redefining the central purpose of the French Radio Project, Olivier's approach and actions fit what identity theorists call "restorative identity work." It involves downplaying one's roots in a stigmatized background in favor of opting into more positive and privileged identities. In Olivier's case, she used her standing as an educator of French to distance herself from her own Cajun heritage while spreading a wider appreciation of "traditional" French culture. "Restorative identity work" is one of several ways educated professionals from stigmatized backgrounds might navigate through the tensions they feel between the respect their profession accords them and the stigma their outsider status often inflicts on them. An individual's background and personal experiences typically shape the varied paths and strategies educated professionals might choose to grapple with their ties to a stigmatized identity. Studies focusing on professions such as educators, medical specialists, and journalists indicate that some seek to use their positions and influence to challenge and ultimately diminish the strength of popular stereotypes associated with a stigmatized identity. Concealing one's connection to a stigmatized background is another notable strategy. In the United States especially, for example, those like Olivier from stigmatized backgrounds who might otherwise pass as "white" can attempt to conceal or evade any connection to or engagement with the stigmatized identity altogether. "Doublethink" is yet another strategy in which such professionals acknowledge the stigma associated with their background but deny its centrality or adverse effects in their own lives.[38]

Hints of each approach are evident in Olivier's efforts and the report she submitted. However, her overall inclination to mask any Cajun connections and instead present herself as a more genteel and proper French-speaking Acadian professional educator nonetheless hewed closest to the concept of restorative identity work. It reflected the influence of her youth growing up in an economically well-off and educated family that nonetheless struggled to communicate in the standard French other elites valued. Olivier's outward presentation of her French roots in the context of the education she received gave her cultural capital; that standing effectively amplified her voice and influence at Harley Smith's expense in the early days of the French Radio Project. It also mattered that Olivier was perceived as a French-speaking bilingual white woman. While the era's ongoing racialization of Mexicans rendered the Spanish they spoke as a potential nonwhite internal threat to the United States, spoken French remained a language popularly associated with supposedly highbrow white European culture. In the context of the Deep South, part of the Acadian cultural revival that Olivier helped spearhead went to great lengths to emphasize a sense of European whiteness in contrast to purportedly ethnic Cajuns.[39] In so doing, Olivier leveraged her upbringing, her French ancestry, and the supposed cultural authority that background gave her. She used her professional position as an educator to legitimize the idea that being of French descent in Louisiana was something different and something better than being an "ethnic" Cajun. In short, the report Olivier submitted to the Rockefeller Foundation was a manifestation of her restorative identity work.

The foundation, however, was not interested in such a report or the self-serving interests it reflected. Foundation officials simply wanted to know if the funding they provided was being used in support of the project's professed objectives. There was no doubt, foundation officers acknowledged, that Olivier's report conveyed her genuine enthusiasm for the work she conducted. However, that work as described was not commensurate with the foundation's rationale to invest in it. Albert Mann, the foundation's vice president and director of the Southern Education project who had visited LSU in the wake of Harley Smith's resignation, was among the several officials who lamented that this effusively self-congratulatory "one-sided" report was clearly disconnected from the project's stated educational purpose. The foundation expected her report to provide substantive evidence demonstrating that the shift to standard French at her behest was warranted. Consequently, the foundation rejected the report and instructed Griffith to provide "a fuller report giving larger insight into the many types of radio programs broadcast as a result of the field study."[40]

By mid-1939 the task of mollifying the foundation fell entirely to Griffith. The now-former university president James Monroe Smith, who had been central to launching the project, was a fugitive. He fled Louisiana in June trying to evade arrest on both state and federal charges of tax fraud, forgery, and embezzlement. Governor Huey Long had appointed the pliant Smith to LSU's presidency in 1931 as part of a larger goal to transform LSU into a nationally competitive university. By the time the controversial Long was assassinated in 1935, at that point serving as a US senator, his efforts resulted in a New Orleans-based university medical center and the development of a formidable athletics program prioritizing football. Along the way, LSU's enrollment increased from 2,100 to 8,550 students between 1930 and 1939. However, with the domineering Long out of the picture, by 1939 Smith had bilked the university out of tens of thousands of dollars. He used state funding to build his family a residential mansion on campus and frequently commandeered the $20,000 university plane purchased for athletic recruitment to facilitate his personal travel. Smith, who was apprehended in Canada and extradited back to the United States in July, was ultimately convicted on multiple state and federal charges and received a lengthy prison term.[41]

With President Smith out of office by June 1939, Griffith attempted to assuage the foundation's concerns by enclosing a one-page single-spaced report on the broadcasts written by Ralph Steetle. This report was also light on the specifics of listener engagement. Instead, it focused primarily on describing how the broadcasting portion of the project functioned, including program development, topics of programs, and cooperating stations. Steetle, who had bluntly expressed his own doubts about the project's viability after Harley Smith's resignation some six months earlier, nonetheless concluded at this juncture that "the results achieved by the French program justify the time and energy expended on it." In support of that assertion, he cited only one anecdotal example. Steetle shared an extension agent's story of being unable to convince area farmers to build chicken breeders. However, after the broadcast of a program on that very topic, farmers supposedly started building the structures and then sought out the agent to inquire if they had been built "like the radio said." For good measure, Griffith included a sample script on the handling, packing, and marketing of spring vegetables, written in standard French. A script alone, however, offered no further insights on the central issue of concern: how listeners meaningfully engaged with the programming they heard.[42]

The Rockefeller Foundation files contain no evidence of its reaction, positive or negative, to LSU's revised submission. Griffith's June 14, 1939, response to the foundation's concerns is the last correspondence in those records. In that letter Griffith expressed a justifiable concern that the foundation might yank the funding.[43] Neither the Rockefeller or LSU records contain a reply or any future reports. The archives at LSU refer to the project as continuing into 1940. Several effusive letters of praise received from participants residing in the various involved parishes through the middle of 1940 are preserved in the files, but there is no indication of foundation support continuing beyond that point.

Even if the Rockefeller Foundation proceeded with a second year of funding, it certainly did not extend funding into a third year. Well before 1940, the project had morphed into something completely different than what was originally pitched. In explaining the French Radio Project for LSU's new president John Herbert in December 1939, Griffith described the initiative's primary purpose as seeking "to create interest in the preservation of the French language," with a secondary purpose being "to improve the type of French used in Louisiana." Only Griffith's subordinate tertiary objective, "to give the French speaking people of the state the information they needed in the language they best understand," had any resemblance to the project's original conception. The fact is, as Harley Smith initially conceived it, "the language they best understand" was supposed to be Cajun. Instead, by 1939 Cajun had been entirely expunged from every aspect of the program.[44]

It was not just the purging of Cajun from the project that disconnected it from its intellectual and methodological origins. By 1940, Louise Olivier was well on her way to removing "radio" itself from the so-called French Radio Project. After 1940, radio listening was no longer a part of her "Assemblée Française" gatherings. Instead, they focused entirely on the multigenerational presentations and sharing of stories, dances, and songs that had been the focal point of her enthusiasm from the beginning. A March 1942 *New York Times* story published under the title "Songs Acadians Sing" underscored the extent of the transformation. The article made no mention of "Cajun" or radio. Nor was there reference to any form of instruction, whether in standard French or the Cajun dialect. The piece only focused on celebrating traditional Acadian music. Louise Olivier was not even identified as the project's fieldworker, but as someone whose role in the project had been to "revive the old French songs and melodies and the dances."[45]

Figure 7. Louise Olivier displaying Acadian Handicrafts some twenty years after the demise of the French Radio Project. This image accompanied a story in the October 10, 1961, issue of the LSU student newspaper *The Daily Reveille*, less than a year before her death in 1962. Courtesy of Special Collections, Louisiana State University, via the Louisiana Digital Library.

By 1942, Louise Olivier had, in fact, found another outlet for her commitment to preserving her vision of elite Acadian cultural traditions. She parlayed her apparent passions for cultural preservation into launching the Acadian Handicrafts Project. LSU remained committed to her efforts even after Rockefeller Foundation funding ended. She began recruiting traditional quilters, sunbonnet makers, weavers, and palmetto fan braiders to produce their wares for displays of Acadian handicrafts that made their way to public libraries, local festivals, and community fairs. Craftswomen dressed in traditional clothing accompanied Olivier to such venues and demonstrated their talents for observers. "Personally, I feel we are outstanding people saturated with relics of a passing culture," she worried in 1943. "The customs and even the language of the Acadians are fast disappearing." Historian Fitzhugh Brundage describes Olivier as "a professional cultural activist who brought specialized training and endless stamina to the crusade" of preserving Acadian language and culture. Though her crusade shifted from radio to the production of physical artifacts representing that cultural ideal, her instincts to value and privilege the archaic over the contemporary remained constant.[46]

Her efforts after abandoning the French Radio Project were ultimately consequential. She helped secure state funding to promote the French language and Acadian culture, which in turn helped fuel and inspire other French cultural revival efforts. She did, to be sure, develop her own reservations about the outcomes of her activities. "All that seems to count consists of noise and glitter," she complained in 1955. "I am so tired of hearing 'a pretty girl on a shiny float' and police escorts and brass bands, etc. that I could scream."[47] She learned what Harley Smith long knew: the perspective and experiences an audience brought to the content with which they engage shaped the meanings and conclusions they drew. Olivier's reservations notwithstanding, by the last decades of the twentieth century the various initiatives she helped inspire contributed to a growing popular awareness both in Louisiana and beyond of a collective and primarily white Acadian identity. Her "success" ultimately obscured the centrality of inextricably intertwined Cajun and Creole influences, both Black and white. We will never know if Louise Olivier could have overcome her scream-worthy concerns about the superficiality of the reactions she observed and be satisfied with what she helped achieve. She died in 1962 at age fifty-six, six months after being diagnosed with cancer. Her obituaries celebrated her as, in the words of one local paper's front-page headline, a "promotor of Acadian folklore," saying nothing of the French Radio Project that launched her career as a cultural activist.[48]

* * *

For all their differences, Harley Smith and Louise Olivier had one significant trait in common: they sought to exploit broadcasting as a vehicle to achieve some wider societal change. To do so, their approaches to script development and language usage in the French Radio Project reflected their respective but contradictory existing values and worldviews. Olivier ultimately succeeded in supplanting Smith's initial vision of Cajun education with one that celebrated standard French and Acadian cultural preservation. Olivier's assumptions, prejudices, and convictions about Cajun may not have been universal, but she accrued the power necessary to shape the broadcasting content that reached larger audiences regardless of their preferences.

With Olivier's ascendancy, LSU's French Radio Project never had a chance of achieving any of its original educational objectives. However, the larger significance of this short-lived and ultimately disappointing educational initiative lay not with understanding the obvious reasons for that failure. Instead, Olivier's coopting of the French Radio Project highlights a potent early- to mid-century vision of the history and identity of Louisiana's French-descended communities. Moreover, Olivier's goals and machinations have relevance beyond their specific historical moment and place. Promoting the use of a standard and supposedly purer form of a language as a tool to ward off or arrest larger cultural decline is a favored tactic of linguistic and cultural purists of many eras and in many places. In the late 1960s, for example, a "Council for the Development of French in Louisiana" (CODOFIL) dedicated itself to ridding the region of "red neck French." To do so, this committee recruited French teachers from France, Belgium, and Quebec to facilitate the widespread learning of standard French. Battles to protect the purity of the French language have long been waged in France itself. Beyond the French language, comparable battles to purportedly protect the linguistic purity of a standard language against crass degradation as a means of protecting history and identity have been waged in different countries and cultures for centuries.[49]

In more recent years, the United States has certainly not been immune to fights waged over the prioritization of so-called Standard American English. In the mid-1990s, a controversial debate over using Ebonics—or Black English—as an educational tool in public schools had echoes of the value-laden language disputes that surrounded the lesser known "French Radio Project." A 1996 plan devised by the Oakland School Board to incorporate Ebonics into its curriculum channeled Harley Smith's motivations to educate a disadvantaged audience in the language they best understood. However, to critics venting on the editorial page of Louisiana's *Alexandria Daily Town Talk* in December 1996, the plan was "harmful to blacks," a "form of defeatism," and promised to produce "social cripples." "Speaking properly has no substitute," wrote nationally syndicated columnist Clarence Page, an African American, in the fourth of four critical editorials published on that page. Drawing the comparison to Cajun, linguist Barbara Birch noted that the most visceral reactions against the Ebonics plan reflected the "pernicious view that people who speak 'ungrammatically' are somehow deficient, stupid, bad, even criminal." Such convictions, she continued, illuminate a longer history of "deeply embedded ethnic, social, economic, and gender privilege" that the preferred standard language helps perpetuate. That assessment rings true to Olivier's insistence more than a half-century earlier that standard French be prioritized over Cajun in the ill-fated French Radio Project.[50]

Olivier's success in transforming the French Radio Project into one that reflected her own vision of Acadian cultural purity is ultimately a testament to her own strength and persistence in the face of formidable obstacles. One might have wagered that Harley Smith had the advantage over Olivier. He was, after all, the LSU doctorate-holding white male with linguistic expertise who devised the project and was seemingly friendly with the university president, its Extension director, and the Rockefeller Foundation officials who advocated for it. Smith, in that light, surely should have occupied the more influential and privileged position over Olivier. However, Olivier gained her advantage over Harley Smith by capitalizing on the very stereotypes of Cajun backwardness and French language superiority that the linguist had hoped to challenge. Harley Smith, for his part, was no worse for the wear after being ousted from his own project. He subsequently forged his own long and influential scholarly career in linguistics at LSU, and his scholarship continued to be cited into the twenty-first century.[51] Though the French Radio Project helped launch both their careers and their feud, both Smith and Olivier went on to make their biggest marks in their later endeavors, while the French Radio Project and the lessons it offered on language and identity slipped into obscurity.

CHAPTER 4

"An Efficient Way to Spread Shakespeare's Beautiful Language"

"Basic English," Language Education, and American International Radio, 1935–1941

The *New York Times* dubbed it "Schoolmaster for the World." In 1936, Walter Lemmon's nearly two-year-old World Wide Broadcasting Foundation and its Boston-based shortwave station W1XAL captured the *Times*'s attention. The organization was dedicated to educational broadcasting, rejected advertising support, and focused on engaging foreign as well as American audiences. "He has found a growing audience in Europe and America for radio courses not only in languages but in literature, art, music, science and world affairs," the paper marveled. It further noted that the WWBF's ties to area colleges and universities, including Harvard, Boston, and Tufts, helped produce quality programming and, by the station's estimation, attract more than half a million regular listeners.[1]

The World Wide Broadcasting Foundation was the culmination of Lemmon's decades-old vision. While serving as a US Navy radio officer during World War I, the inventor, electrical engineer, and business executive was first inspired to create a "radio university" to benefit humanity. By 1927, he had earned a small fortune from his profitable radio inventions and had launched the WWBF's predecessor organization, the Short-Wave Broadcasting Corporation of New York, as a commercial-free vehicle to carry "good will programs." By 1929, he moved that organization to Boston, which offered a better location for a transmitter suited to far-reaching international broadcasts. He rechristened it the World Wide Broadcasting Foundation in 1935 and articulated a more fully formed mission prioritizing educational and other highbrow programming. The WWBF's nonprofit mission remained tethered to encouraging "the spirit of international understanding and enlightenment" throughout the 1930s. It was especially committed to "strengthening further the spirit of solidarity and harmony among the nations of the Western Hemisphere."[2]

Lemmon identified language education—specifically the teaching of English to non-English speakers—as central to these ambitions. "A common language tie is the first and greatest step to understanding," Lemmon proclaimed. In his view, the "permanent value" of such a radio-based initiative extended beyond the fostering of international understanding. Its success promised to "have a future bearing on business as well as cultural relations." Lemmon's conviction on this point was long-standing. "Trade aid is also object," proclaimed the subtitle to a 1931 *New York Times* article reporting on Lemmon's plans and his belief that such broadcasts "pave the way for greater acceptance of American products abroad." Some two decades later he still contended that "every additional thousand people in Latin America who become familiar with the English language as a secondary tongue increases the opportunities for closer inter-American understanding and for future trade with us."[3]

To achieve this goal through broadcasting, the WWBF sought to teach a simplified version of the English language. The system, known as "Basic English," was supposedly easier for non-English speakers to learn. Finally, Lemmon and his educational and broadcasting partners did not simply presume that saturating the airwaves with English language lessons would be singularly sufficient to achieve their goals. Reflective of the most up-to-date radio and pedagogical research, the WWBF developed strategies and programs intended to meaningfully engage their prospective radio pupils and maximize the potential of that listening audience's ability to learn English.

In the end, though, the Basic English broadcasting initiative failed to gain the traction, achieve the influence, or produce the benefits Lemmon envisioned. Culturally rooted and period-specific convictions about American and Anglo-Saxon exceptionalism blinded Lemmon and his associates to the key shortcomings of their strategies. These blind spots left the Basic English broadcasting initiative poorly suited to engaging a broad Latin American audience, despite genuine efforts to do so. Basic English was itself weighed down by cultural and ethnocentric baggage specific to its historical moment. Its appeal as a vehicle to promote international exchange and understanding says more about the exceptionalist visions of its adherents than it says about the WWBF's ability to facilitate the learning of the English language across the globe.

A World Radio University

For all the popular enthusiasm expressed for radio's supposed potential to encourage international peace and understanding, one obstacle loomed especially large: language barriers. They were obstacles that prevented programs in one language from engaging listeners who only understood another. For the most part, early international broadcasting initiatives put little thought into surmounting such barriers. Though commercial broadcasters like NBC and CBS developed some programs deliberately designed to increase American understandings of Latin America, when looking to audiences abroad they often adopted the most inexpensive practice of simply rebroadcasting programs initially produced for US domestic audiences. The divide between the rhetoric of seeking mutual understanding and the substance of such broadcasts was glaring. In the mid- to late 1930s, for example, China-based US officials successfully lobbied for the establishment of a shortwave station that could reach East Asia with English language broadcasts, claiming such programs promoted trade, peace, and understanding. One trade official making the case for the new station presumed there was a Chinese audience of some eighty million people fluent in English, many of which had attended US colleges and therefore represented an influential cross-section of the Chinese population that the United States should be most interested in engaging. The resulting station, a General Electric-owned station that carried NBC network programming, did not capitalize on this imagined audience. Not surprisingly, it instead appealed primarily to Americans residing in East Asia seeking the familiarity of programs from their homeland.[4]

Walter Lemmon wanted to be the exception. He was determined to surmount language and cultural barriers. The result, he hoped, would be a genuine realization of international peace and unity spread through broadcasting. His pedigree helps contextualize these convictions. He earned a degree in electrical engineering from Columbia University in 1917 and was subsequently commissioned as a lieutenant in the Navy during World War I. By 1919, the twenty-three-year-old Lemmon was serving as the radio officer and scientific aide aboard the *USS George Washington*. It was an important assignment. The ship was taking President Woodrow Wilson to the Versailles Peace Conference, where Wilson hoped to facilitate a settlement that reflected his internationalist visions of peace and cooperation following the destructive war. During that trip Lemmon, a longtime radio enthusiast, tinkerer, and inventor, developed the radiotelephone system that allowed Wilson to communicate directly with Washington during the journey. Each afternoon, Lemmon used that same system to entertain Americans on the ships in the surrounding ocean and those stationed in army camps in France. To do so, he broadcast phonograph records and invited soloists to the radio cabin to sing directly into the microphone. The technology fascinated Wilson. The president often dropped by to visit with Lemmon and learn more about the medium. As Lemmon recalled in a 1924 interview he gave shortly after Wilson's death, the president came away from his discussions convinced—like many of his contemporaries soon would be—that "radio would become the greatest medium for linking of thought that the world had ever known."[5]

To Lemmon, the need to deliberately deploy radio in just that way became painfully apparent at the Versailles Conference. During those strained negotiations, he saw firsthand how labored and tedious communications were against the backdrop of language barriers and cultural divisions. In Lemmon's view, translations could not adequately surmount such obstacles. Wilson had traveled to Paris hoping that internationalist initiatives and organizations such as the League of Nations could provide an institutional framework for nations to engage each other and maintain long-term peace, prosperity, and understanding. Lemmon, however, returned from that trip with the conviction that international radio was necessary to help establish the intellectual and cultural common ground required to achieve that Wilsonian vision. On the return journey from France, Lemmon shared with Wilson his idea for an international radio university to help build such a foundation. Wilson, a former president of Princeton University who was already excited about radio's potential, was predisposed to embracing such educational aspirations. He pledged his support.[6]

The timing, however, was not right. Debilitated by a devastating stroke soon after Versailles, Wilson never could deliver on his promise. The US Senate's subsequent rejection of the Treaty of Versailles raised fundamental questions about

Figure 8. Walter Lemmon. Courtesy of the Rockefeller Archive Center, Sleepy Hollow, NY.

the broader American commitment to the Wilsonian internationalism that also girded Lemmon's radio visions. Beyond the unfavorable political climate, an even greater obstacle to executing Lemmon's vision of a radio university was the unsatisfactory technological context of the early 1920s. Broadcasting was only in its most nascent stage, and the technology needed for reliable and effective long-distance international broadcasts had not yet been perfected. It was only by the end of the 1920s that shortwave radio technology and the ability to reliably send broadcasts to points many thousands of miles away had been developed sufficiently to support the pursuit of Lemmon's vision of wireless internationalism.[7]

In the interim, Lemmon bided his time. Shortly after the war, he became a founding partner in the New York-based Malone-Lemmon Radio Laboratories, which manufactured radios. Amidst the broadcasting boom that transfixed the country by the late 1920s, Lemmon organized the Radio Industries Corporation in 1929 to pursue the development of patents in the field. Befitting his skills as a radio tinkerer and inventor, he patented the "Single Tuning Dial Control" that became the standard for radios. He also invented a radio-operated electric typewriter. The money he earned from those two inventions assured his financial security going into the 1930s, with radio being one of the few industries that grew and thrived during the Great Depression. In 1929, RCA acquired the tuning patent, while in 1932 IBM bought the typewriter patent. In addition, the now-millionaire Lemmon joined IBM's executive ranks as general manager of the radio-type division. By the end of the 1920s, Lemmon had become a well-connected and wealthy man in his own right, which positioned him to launch what became the World Wide Broadcasting Foundation. The WWBF would put into practice Lemmon's vision of a radio university first articulated to Woodrow Wilson aboard the *USS George Washington* a decade earlier. The WWBF was a radio organization "dedicated to enlightenment," the foundation's motto professed. It was "devoted entirely to educational and cultural programs," underscored a 1937 *New York Times* article. And it was equipped to "aid world wide cooperation," as Lemmon himself was quoted as saying in the March 1937 issue of the *Scientific Monthly*.[8]

Lemmon's vision for the WWBF reflected his effort to address a popular critique of the commercialized American system of broadcasting: American radio should emulate the supposedly more high-minded and highbrow approach of the advertising-free British Broadcasting Corporation. By the 1930s, that view particularly resonated among those dispirited by the marginalization of educational broadcasting stations on the American radio dial. Shortwave broadcasting, however, offered a means to pursue such lofty ambitions outside of the heavily commercialized AM radio dial most familiar to Americans. Because the

US government designated shortwaves as an experimental frequency range, that spectrum seemingly offered a hospitable space for Lemmon to pursue his ideas for an alternative to the commercialized American system of broadcasting. As shortwave was an experimental medium, the US government granted shortwave licenses with a stipulation that prohibited the sale of advertising to support programming. The broadcast of programs over these stations was instead expected to help provide information and data that increased understanding of how that portion of the frequency spectrum worked. Granted such a license amidst the economic challenges of the Great Depression, Lemmon invested $100,000 of his own money to build Boston-based shortwave station W1XAL in 1931 (renamed WRUL in 1939) as the heart of his enterprise. In subsequent years, Lemmon expended up to $11,000 annually from his personal funds to sustain the station's educational broadcasting initiatives. In 2023 dollars, Lemmon's investments amounted to an initial outlay of nearly $2 million and annual expenses exceeding $235,000.[9]

It was that financial burden that led Lemmon to the philanthropic Rockefeller Foundation to seek additional funding. However, Lemmon was not merely looking for money to defer his own considerable expenses. Rather, he looked to the Rockefeller Foundation for funding that could help ensure the WWBF's programs effectively engaged its target audiences. Lemmon favorably distinguished himself from many of his contemporaries, who often celebrated the promise of radio to bring diverse peoples together in common cause with little more than superficial rhetoric. The WWBF founder understood he had to consider how to engage an audience in order to achieve his educational goals. Toward that end, he asked for Rockefeller money to improve the effectiveness of his enterprise. Seeking to hire qualified staff writers, produce experimental educational programs, and—perhaps most importantly—fund studies checking the effectiveness of those programs, Lemmon rather ambitiously requested $25,000 in 1935, followed by subsequent requests for $40,000 and $100,000. Harkening back to the key lesson he learned from observing the negotiations at Versailles nearly two decades earlier, his funding requests emphasized the development of programs to support language study in a way that sustained listener interest and ultimately improved the lives of those who listened.[10]

The Rockefeller Foundation was a good fit for Lemmon's WWBF. First, as was the case with LSU's contemporaneous French Radio Project, Lemmon's requests for funding coincided with the foundation's decision to prioritize support for both radio and language education projects. Second, Lemmon's idea for an international initiative to improve the teaching and learning of English to non–English speakers meshed with the foundation's corollary interest in advancing understandings of how radio engaged and influenced listeners.

Moreover, an international educational radio initiative as Lemmon envisioned was complementary to the foundation's larger quest seeking to remedy the causes of the multiplying social problems that fueled poverty, disorder, and instability, and in so doing "promote the well-being of man-kind throughout the world." In pursuit of that overall objective, the Rockefeller Foundation, in fact, directed much of its philanthropy toward Asia, Africa, and Latin America. These regions were deemed the undeveloped and backward parts of the world. They had supposedly been left behind as the West modernized and industrialized. Internationally focused Rockefeller projects were often premised on the conviction that the millions of "uncivilized" beneficiaries of these various projects stood to benefit from an embrace of advanced American "civilization." If these beneficiaries spoke English, that task could be made much easier. Such aspirations reflected the American exceptionalism deeply embedded in US foreign relations: a desire to "spread the American dream" based on the conviction that what was good for American capitalism would benefit all peoples who encountered it.[11]

Lemmon was of a like mind. Such inclinations infused his speculations that listeners across Latin America would benefit from his broadcasts. He believed that the cumulative effect of the WWBF's cultural and educational programming should "properly emphasize the accomplishments of the American free enterprise system." But these accomplishments, in Lemmon's eyes, were not solely about finding more markets for American businesses; by following the Americans' example, Latin American nations and peoples could be positioned to follow a more productive path. His broadcasts, he believed, could dissuade Latin American listeners from believing "the false and often planted impressions that the United States is interested only in materialistic progress and entirely devoid of cultural interests and of the humanities." Moreover, this model, Lemmon assumed, was within reach of the peoples of Latin America, in large part because of the common ground that already existed between their respective countries and the United States. To convey that point and to entice the peoples of Latin America to embrace their northern neighbor more emphatically, Lemmon envisioned developing programs that showcased notable accomplishments in American literature, art, architecture, and music in a way that underscored how Spanish and Latin American traditions infused some of the best accomplishments of American culture. To be sure, Lemmon was mindful of the possibility that his station's programming could inspire anti-American opposition. However, he believed that the sincere intentions behind his efforts would be rewarded. "Since World Wide takes no commercials, it is not resented and Latin American stations avail themselves of the privilege of picking up and rebroadcasting any of World Wide's programs," the *New York World Telegram* reported as fact after interviewing Lemmon.[12]

The Rockefeller Foundation, excited by the WWBF project and its prospects, was inclined to fund Lemmon's proposals, albeit not at the robust levels he had first requested. As was the case with LSU's French Radio Project, the WWBF benefited from Assistant Director of Humanities John Marshall's enthusiasm for radio education projects. The foundation's favorable internal evaluations of the WWBF's proposal pointed to an overall excitement for educational radio projects in general and the specific common ground the foundation shared with Lemmon's vision of educational radio for international audiences. It singled out the WWBF's emphasis on programs that sought to inspire educational and cultural uplift, with ambitions for English language instruction earning special mention. The WWBF's ties to the prestigious universities of Boston and the surrounding area further bolstered the Rockefeller Foundation's confidence in the substance of the WWBF's initiatives. In 1935, the foundation approvingly noted there was "direct cultural interchange with listeners in other countries." It cited as evidence the combination of rapidly increasing shortwave receiver ownership in both Europe and Latin America and the praise-filled listener letters that arrived from those areas of the world. The WWBF's most substantive contribution appeared to be its efforts to convey "aspects of American life which seldom figure in the news, in motion pictures, or in the network programs which are ordinarily re-broadcast on short-wave." It was a vital attribute, the foundation concluded in its approval of a 1938 funding request, "at a time when shortwave broadcasting is increasingly an instrument of national projection."[13]

The WWBF's educational aspirations did not preclude its own inclination to position itself as "an instrument of national projection." In the WWBF's own telling of its brief history, it touted the existence of thousands of listeners across Europe and in other "more remote sections" of the world who sought out the WWBF's programs. The organization therefore claimed it occupied an ideal position to "present American education, culture, and ideals to many other countries in such a manner as to cultivate respect and cooperation." Through this positioning, it expected to play an important role in "improving international good-will directly with the citizens of other countries." Moreover, English language education occupied a central place in the WWBF's strategies to promote international goodwill through the spread of American ideals and values. Given the vast transmission range of shortwave broadcasts, Lemmon had long imagined a large pool of potential language students across Latin America and extending to every remote corner of those countries that otherwise offered "almost no local opportunities for cultural progress."[14]

Lemmon also favorably impressed the foundation by taking advantage of additional resources he had at his disposal. The WWBF cultivated connections with experts at the Boston area universities to devise an English language education program suited to thrive over the radio. Lemmon also took advantage of his access to the foundation-funded Radio Research Project to analyze how the WWBF's programs engaged its audience. By 1938, the Radio Research Project, led by sociologist and media theorist Paul Lazarsfeld, was at the forefront of scholarly research in reception studies. Lazarsfeld and his cohort explored how and why listeners engaged with the radio programs they encountered. Their ongoing research and conclusions, much of it published two years later in Lazarsfeld's milestone book *Radio and the Printed Page*, concluded that "radio was a powerful social institution" and "the preferred medium of the more suggestible man" likely to be found among "people on the lower cultural levels." Lazarsfeld was no technological determinist, however. He concluded that radio, in fact, was not totalizing in its influence. Instead, the evidence pointed to how listeners selectively absorbed and filtered content based on factors such as their own position in their larger social framework and related worldview, as well as how radio itself fit into the larger context of media with which they engaged. Such conclusions, in fact, were an early expression of now-standard social constructivist approaches to understanding media influence and effects.[15]

For his part, Lazarsfeld was favorably impressed by this initiative. He saw great potential for it to engage an audience on its own terms. Lazarsfeld discerned especially high listener enthusiasm for WWBF's broadcasts among an impressively large audience (albeit while admitting that it was difficult to ascertain how large that audience really was). There was, he believed, a "great reservoir of people easily in reach" and that shortwave radio was "perfectly suited" to the task. However, Lazarsfeld cautioned that there was still room for improvement, especially amidst some signs of a possible listener drop-off. "The great enthusiasm of the listeners for the station could be used more systematically for the expansion of the audience," Lazarsfeld advised.[16]

Lemmon took this advice to heart. In 1939, he believed he had found the ideal approach to systematically execute his objectives for radio-based language education. "Basic English" was a relatively new and greatly simplified version of the English language, seemingly tailor made for radio. One of the linguists closely involved in its creation, I.A. Richards, had recently arrived at Harvard and was collaborating with the WWBF to adapt the Basic system for instructional broadcasts. With war clouds ominously gathering over Europe by 1939, the WWBF's efforts to teach Basic English confronted an international climate in which the need for such understanding and cooperation seemed greater than ever. As far as Lemmon was concerned, there was no better moment than the present and no better tool than Basic English to realize his now two-decade-old vision of educational radio as a vehicle to promote international peace and understanding. The Rockefeller Foundation agreed, and between 1938 and 1940 provided the WWBF with some $10,000 specifically earmarked for developing and broadcasting Basic English programs.

Basic Broadcasting

The WWBF began its Basic English broadcasts in September 1939. By this point, the Second World War was spiraling to ever larger proportions in both Europe and Asia. However, it was the previous world war that loomed even larger in bringing this radio project to fruition. Just as World War I inspired Walter Lemmon to envision what became the WWBF, that same conflict propelled linguist C.K. Ogden, with assistance from I.A. Richards, to create a simplified version of the English language. While Lemmon believed communication breakdowns plagued the Versailles Peace Conference seeking to conclude the war, Ogden and Richards assumed that the so-called Great War itself was the result of a complete collapse of rational international communication in the period leading up to its outbreak in 1914. With that inspiration, Ogden and Richards worked on developing the Basic English system throughout the 1920s to facilitate international communications. The system was unveiled in 1930. Echoing Lemmon's rationale for the WWBF's global value, Richards insisted that Basic English was ideally suited to foster international understanding in a variety of scholarly and practical topics, including history, culture, and business. Fluent English speakers, beginning students, and anyone in between would find in the simplified Basic English system "a medium which conveys to them the living thought of the world." It was, Richards argued, "a means of helping planetary purposes as opposed to sectional aims, of easing tensions due to ignorance, of making the interdependence of peoples more manifest, of widening and enriching mutual awareness, and of consolidating world opinion against disruptive strains." In C.K. Ogden's eyes, the radio and Basic were a perfect match. The pairing of Basic and radio were the essential means by which "the future may be guided to peace." That connection was why, in Ogden's view "our first hope for the future of Basic English is the short-wave radio."[17]

To accomplish that ambitious goal, Basic English attempted to strip the language down to only its barest essentials. Basic English is comprised of a mere 850 commonly used words. The overall Basic English vocabulary list is broken down into 650 nouns, 150 adjectives, and 100 so-called operative words (such as *if*, *the*, and *very*). Just 16 verbs can be found in the operative word list (such as *come*, *do*, and *take*). Replacing some 4,000 verbs in the unabridged language, those 16 action words were deemed sufficient to convey far more than their 16 basic definitions; they could be combined with other words from the overall list to convey meanings synonymous with many of the excluded English-language verbs. For example, students of Basic English were not taught the verb *continue*, but rather were taught to combine two words from the master list—*go on*—in order to convey the same meaning. Clarity was presumably enhanced through the sacrifice of unnecessary detail, jargon, and bombast while the variety of word combinations used in the system could still ostensibly convey sophisticated ideas and concepts. In the process, proponents of Basic argued that the richness of culture and history, as well as economic, scientific, and technological knowledge, could be shared by peoples of different national and language backgrounds who spoke this simplified version of English. Ultimately, Basic English's advocates were convinced that the system offered a relatively easy way to learn the language quickly and communicate effectively with native speakers in a manner that did not necessarily have to sacrifice the richness and complexity of ideas embedded within the full language.[18]

Basic English was, first and foremost, a product of its time. Ogden and Richards readily pointed to World War I as the source of their inspiration. In fact, Basic was just one of many transnational language and communications initiatives dating back to the nineteenth century seeking to foster internationalism and cross-border cooperation. Increasing numbers of people in different parts of the world expressed concern over intensifying nationalism and the growing number of international conflicts it seemed to fuel. In this view, World War I was the unfortunate culmination of such longer-standing concerns. With language divisions often blamed for that nationalism and the resulting conflicts, the pursuit of humanistic language reform, many surmised, might provide the mechanisms that helped create common meanings and understandings shared across national borders. The creation of the international language of "Esperanto" in the 1880s is perhaps one of the most well-known of these initiatives that reflected such anxieties. The development of Braille in the late nineteenth century, which also had transnational reach, was another such initiative that helped satiate such concerns. Alongside these language-specific initiatives, photography and motion pictures' ability to convey a message visually regardless of a person's native language also generated enthusiasm in the late nineteenth and early twentieth centuries; they too could be powerful border-crossing vehicles of communication that might further international understanding. Such widespread convictions helped make the pairing of the WWBF with Basic English so incredibly appealing.[19]

For its part, the Rockefeller Foundation had been enthused about Basic English's utopian promise well before it gave its first dollar to the WWBF. Prior to funding Lemmon's outfit, the foundation awarded a grant to Ogden and Richards's Orthological Institute in London to study the usefulness of teaching Basic English in China and Japan. The philanthropic organization justified its support of Basic on the basis that it promised to "improve relations in the Far East" with the ultimate purpose of "promoting a fuller understanding of Western civilization in the Far East." Richards, who by 1939 taught at Harvard and helped introduce his Basic system to the WWBF, concurred. In June of that year, with an eye on the two-year-old Sino-Japanese War that wracked East Asia, Richards lectured that any chance of fostering a "clear understanding" among the peoples of Asia, as well as between the peoples of Asia and the West, depended on improving the techniques for learning English. For the benefit of his American audience to whom he spoke, Richards also underscored how Basic could promote "good neighbourly" US-Latin American relations, a priority of President Franklin Roosevelt's administration since it took office in 1933. Also throwing his weight and reputation behind Richards was the influential scientist Sir Richard A.S. Paget, whose research into language, communications, and comprehension (most notably regarding deafness and sign language) remained influential throughout the twentieth century. Basic English, Paget insisted in response to Richards's lecture, represented "the beginning of a great movement... of great importance to mankind." Given it supposedly took only two weeks to secure a working knowledge of Basic, one unidentified audience member marveled that the cost for achieving this vision was "not, after all, a very high price to give for the key to good international feeling."[20]

That English was the logical choice to facilitate such exchanges also seemed obvious to Richards. Given the globally pervasive knowledge of English, Richards believed the Basic system put the goal of creating a true "world language" within reach. To be sure, Richards, an Englishman, likely did not share Lemmon's specific enthusiasm for promoting particularly American ideals. However, the ties binding England and the United States offered substantial common ground between the ideological outlooks that motivated both Richards and Lemmon. By the twentieth century, educated elite opinion in both countries often celebrated the deep cultural and intellectual ties that purportedly bound the two nations together representing the "Anglo-Saxon" vanguard of the technologically and economically advanced civilized world.[21]

Perhaps Richards quibbled with assertions claiming some unique American superiority. However, he undoubtedly identified the United States as one of the instrumental actors in a global English-speaking Anglo-Saxon panorama purveying the essential values and advantages of the so-called civilized world. Richards also had confidence that the groundwork for Basic's expansion was already in place. "It is the mother tongue or the administrative language of over

600 million people," he lectured in 1939. "It is the official second language in most of the great educational systems. It is the trade language of the Pacific. It has become all this without any special efforts from us to make it so." Ogden echoed many of the same talking points. Both men envisioned a combination of native and non-native speakers learning Basic throughout the world. Collectively, these students would soon be well equipped to communicate with each other regardless of their national backgrounds or native tongue. They had powerful company in these convictions. British prime minister Winston Churchill certainly believed as much in 1944 when he encouraged US president Franklin Roosevelt to put the full support of the American government behind "the promotion of Basic English as a means of international intercourse." The simplified language would, Churchill continued, "prove to be a boon to mankind in the future and powerful support to the influence of the Anglo-Saxon peoples in world affairs." Roosevelt agreed that the idea had "tremendous merit" and asked his secretary of state Cordell Hull to drum up support in Congress for such a global initiative. "I really believe," Roosevelt wrote, "that the other nations would go along with us." Richards concurred, and pointed to the fact that Basic English education programs were already underway in India, Burma, Greece, Denmark, and China.[22]

Pin Pin T'an, Richards's former student from China, became the voice of the WWBF's Basic English broadcasts to Latin America. The foundation had been funding Ogden and Richards's Basic English programs at China's Yenching University since 1933. Richards himself reached China in 1936 with the support of additional Rockefeller funding to establish the Orthological Institute of China at that school. The goal was to further improve the teaching and learning of Basic English at the institution. It was at Yenching that Pin Pin T'an first met and studied Basic under I.A. Richards. However, T'an's 1937 graduation from Yenching with a degree in English coincided with the outbreak of the Sino-Japanese War. That war's first battle, and by extension the first battle of World War II in Asia, was fought just a few miles from the university. As the conflict unfolded around the university, T'an left China to pursue graduate study in the United States. She secured a graduate fellowship from Radcliffe College, an elite Harvard-affiliated all-women's institution. Richards, forced to leave China by 1939 amidst the expanding war, also landed at Harvard, where he was reunited with T'an. Even before arriving in Massachusetts, Richards began lobbying the Rockefeller Foundation to enlist T'an in the WWBF Basic English endeavor. The effusive praise for T'an's abilities coming from the cofounder of the Basic English system was a compelling endorsement. By the time Richards arrived on campus in October 1939, the foundation had agreed to fund T'an's salary at a rate of $1,200 per year (the equivalent of more than $26,000 annually in 2023 dollars).[23]

Figure 9. Pin Pin T'an. Courtesy of the Rockefeller Archive Center, Sleepy Hollow, NY.

T'an's enthusiasm for Basic English rivaled that of her mentor Richards and her new boss Walter Lemmon. T'an, a native Chinese speaker, used the Basic system to learn the English language, earn a graduate degree in English at Radcliffe, and then secure a broadcasting job teaching English to non–English speakers. That record of accomplishment was quite the endorsement for Basic's potential. More importantly, she shared with Lemmon the conviction that strategies to teach Basic over the radio needed to be thoughtful, deliberate, and based on carefully designed approaches. As such, initial expectations for the Basic broadcasts were low. "The Basic series was started very quietly, as an experiment," T'an recounted after nearly a year of broadcasts. In fact, Latin America was not originally the primary focus of the initiative. The early broadcasts instead hoped to attract native English speakers as well as non-native ones in the United States and Europe. The purported value of teaching Basic to native English speakers was, as T'an put it, "to test and clarify their language habits." During the first month on the air, T'an gave daily fifteen-minute news broadcasts in Basic directed toward both European and American audiences to illustrate the system's uses and accessibility. Over the next few months, WRUL added several additional hours of Basic programming per week. One such program was "Basic for Every Field," which included talks by Richards himself.

Foundation officials were pleased with their investment, in large part because of T'an, who received praise for being "unexpectedly resourceful" and doing an "excellent job." The production of the additional programming, however, proved impossible to sustain and the number of weekly broadcasts was soon scaled back in order to ensure quality.[24]

By spring 1940, these efforts attracted favorable attention in Washington. In April, T'an traveled to Washington, D.C. and met with officials from the Pan American Union and various divisions of the State Department. She hoped she could secure official support for the Basic radio project. While noting that all officials she met with seemed deeply interested in her work, she credited Frances Colt de Wolf, chief of the State Department's Telecommunications Division, for the idea of specifically targeting Latin America with Basic English programs. "He said that anything WRUL may do along that line will not only help create better Inter-America understanding, but will stimulate trade and travel," T'an recalled. A closer and more systematic focus on Spanish-speaking Latin America, de Wolf suggested, could offer a better means of exploiting Basic's value as a tool to facilitate international communications and exchange.[25]

T'an found the idea intriguing, but still needed to be convinced of its practicality. To be sure, de Wolf's idea matched T'an's own view of Basic's value as a tool to foster internationalism. Nonetheless, she had well-founded reservations. Echoing Lemmon's approach and demonstrating her familiarity with new communications studies theories, she understood that stand-alone language education broadcasts were not necessarily well suited to achieving their objectives. She argued that there needed to be a complementary institutional structure and context into which the Basic broadcasts would fit, and from within which listeners would be able to engage and make sense of the programming they heard. But there was not a single institute for Basic English instruction in all of Latin America and the system itself was largely unknown there. De Wolf promptly assuaged those concerns. He offered the full support of the State Department, which would send materials to its consular offices throughout Latin America. The consular offices, in turn, would distribute the materials and publicize the broadcasts. T'an was persuaded, and upon her return to Boston in April 1940 she convinced Richards and the project's lead scriptwriter Charlotte Tyler to support the idea. They immediately got to work on adapting the system to engage a Latin American audience and hired a Spanish-speaking broadcaster to work with T'an. By the end of May, the WWBF was broadcasting specially designed Basic programs to Latin America. The State Department followed through on its commitment and sent publicity materials throughout the region.[26]

Despite the hurried development of the Latin American program, it had all the appearances of being an immediate success. After only about a month of broadcasts, a surprisingly large volume of mail from across Latin America was arriving at the WWBF. T'an credited the State Department's cooperation and the efforts of US consular officers across Latin America for publicizing the program and building this audience, just as de Wolf had promised. Still, T'an would not go so far as to credit consular efforts for creating an audience out of whole cloth, but rather for drawing attention to programs for which there was already inherent interest. "Unquestionably, Basic has much to offer those who wish to learn English quickly and easily, and the Spanish-speaking public has not been slow to recognize this."[27]

By the summer of 1940, the Basic English broadcasts to Latin America appeared on solid footing. It had not even been a year since T'an took the first tentative steps toward developing Basic English instructional programs for broadcast and only three months since she expressed reservations about meaningfully engaging an expansive audience in Latin America. The landscape had changed quickly, though. "WRUL's Basic program has, I feel, gone beyond the experimental stage," she professed in a progress report to the foundation. She saw in WRUL the potential to "become the centre of Basic English work throughout the world" if it could be fully funded. However, given funding limitations, she conceded, it would be most effective to prioritize the development of broadcasts to Latin America rather than overextend themselves once more. The Basic radio program "has something very definite to offer to Latin America and the general broadcasts are creating wider and wider interest in Basic," she confidently proclaimed.[28]

T'an used the flood of complimentary letters received from across Latin America to try to secure additional funding from the Rockefeller Foundation. Touting that high volume of mail, T'an argued that further expanding these efforts "would be most rewarding." She envisioned collaborating with the press throughout Latin America to further increase awareness of the project. T'an wanted to establish Basic centers throughout the region. Ideally, they could serve as distribution centers for printed materials, as well as a place where students could go for consultation. She promoted the suggestion of a Brazilian correspondent who wanted to see relationships forged with local schools for those, like himself, who did not own a shortwave radio. T'an was also excited by a Cuban listener's suggestion that arrangements be made for the local rebroadcast of the lessons. Doing so would expand access to the programs by not making their reception contingent on shortwave receiver ownership. "Both suggestions should be feasible and profitable."[29]

The Rockefeller Foundation ultimately agreed. The foundation's John Marshall, a radio education enthusiast, admitted to being compromised in terms of evaluating the effectiveness of the WWBF's instructional efforts. He confessed that his lack of proficiency in Spanish or Portuguese put him at a disadvantage in gauging the overall effectiveness of the WWBF's Latin America project. But with C.K. Ogden adding his own enthusiastic endorsement of the project and its potential, Marshall remained strongly supportive. The foundation awarded the WWBF an additional $2,000 in August 1940 on top of the approximately $10,000 it had received to that point. Additional salary support of $1,200 was designated for T'an so she could stay on and continue developing the broadcasts for another year. The remaining $800 was designated to cover the increasing costs of handling listener mail, as well as travel and other multiplying expenses. The apparent "growing interest in the study of English in Latin America," the foundation concluded, justified the expenditure of these additional funds.[30]

Basic Listeners

The mail arriving from Latin America at the WWBF continued unabated. From September 1939 to July 1940, the WWBF received some 299 letters from listeners throughout the world. One-third of those letters admittedly originated from within the United States (not necessarily a discouraging fact, given the corollary interest in familiarizing native English speakers with Basic). The letters from Europe predictably tapered off once war erupted on the continent in 1939. However, 139 letters came from non-English-speaking countries, mostly in Latin America, which had only been a specific target of the broadcasts since spring 1940. Even more impressive, nearly every country in Latin America was represented in this correspondence. After the foundation reaffirmed its funding for the broadcasts in the summer of 1940, another 301 letters arrived from Latin America in the six-month period spanning September 1940 through February 1941. Of these, 112 came from Central America and the Caribbean, with 189 arriving from South America, including the most southernly distant countries of Argentina and Chile. By the end of 1941, the WWBF claimed nearly 10,000 students had formally "enrolled" in the lessons (although the report making that claim did not clarify how students formally enrolled or how the WWBF arrived at this figure).[31]

The WWBF and the Rockefeller Foundation recognized the problems inherent in relying solely on listener letters to determine the initiative's popularity and effectiveness. They understood that, in real terms, 10,000 supposedly dedicated students only accounted for a miniscule portion of all Latin Americans.

Both the foundation and the WWBF also acknowledged that the broadcasts tended to reach an atypical and "much more cultivated group who frequently express their appreciation of the unusual character of the program's broadcast." The letters unsurprisingly came overwhelmingly from urban areas, where education and literacy were the highest. The foundation's John Marshall was not troubled by these limitations. He rationalized that the broadcasts had reached the best possible cross-section of elite listeners who were "pretty influential in Latin America."[32]

Lemmon and Marshall both saw the large volume of letters received from Latin America as proof that the Basic broadcasting initiative to Latin America was an obvious and unqualified success. Writing in October 1940, Lemmon assured Marshall that the foundation's confidence and funding had been well placed. In Lemmon's view all signs continued to point to an engaged Latin American audience "very eager to learn our tongue." Noting that there were more than a million shortwave sets in use across Latin America, Lemmon was confident that the potential to further grow the audience throughout the region was very high. Marshall, who had been unfailingly supportive of the project from the beginning, was especially surprised by the apparent reach of the Basic broadcasts into the most southernly countries of South America. "These programs are catching if anything more interest than one might have expected," he marveled after a February 1941 meeting with T'an.[33]

The listener response from Latin America did indeed appear very enthusiastic. Letters frequently professed that the broadcasts satiated long-standing desires to learn English, celebrated the potential to further cultural exchanges between the nations of Latin America and the United States, and effused over the global prominence of the English language. A listener from the Dominican Republic claimed a long-standing interest in learning "that beautiful and interesting language." The broadcasts were, he claimed, of "indispensable" value in furthering "firm cultural and intellectual relations between the diverse peoples of Latin America." In fact, more than one-third of the letters T'an sent to the Rockefeller Foundation underscored some manifestation of the claim that Basic was a valuable tool to promote such exchanges. It was "a language so vital to the world," wrote one listener from Corinto, Nicaragua. Another listener from Camaguey, Cuba, gushed over the "marvelous" lessons that were "so useful to all Latin Americans." A listener from Havana, Cuba, went even further and touted the Basic broadcasts as "an efficient way to spread Shakespeare's beautiful language."[34]

T'an's skills as a broadcaster and effectiveness in engaging her audience as an educator also made an impression. Letters frequently praised her pleasant voice, clear diction, and beautiful pronunciation. "My congratulations to the lady who speaks English," wrote one Argentinian listener. "Without flattering her, I compare her pronunciation with the Linguaphone Records." It was high praise indeed to compare T'an to the renowned distance education company that distributed its language lessons via phonograph recordings. The intimate style that infused the broadcasts, noted a correspondent from the Costa Rican capital of San Jose, makes the listener feel as though they are the sole focus of lesson, and consequently "there is no opportunity to lose interest or let one's attention be distracted, but quite the contrary."[35]

A radio program's ability to foster a sense of intimacy and personal connection with a listener was crucial to the popularity many radio programs enjoyed in many different countries. Listeners typically understood they were merely one person within a vast audience of individuals mostly unknown to each other but who were listening simultaneously across a larger geographic expanse. However, deliberate performance styles and tactics on the part of broadcasters and program hosts like T'an sought to engage listeners on a personal level in a manner that transcended feelings of individual anonymity. To reciprocate, many listeners adapted the practice of writing letters to a station or broadcaster. In so doing, they surmounted their own anonymity within that larger sea of unknown listeners by personally identifying themselves to the radio performers that had succeeded in provoking a sense of intimate and personal connection. For example, US president Franklin Roosevelt's "Fireside Chats" inevitably triggered a flood of letters responding to, as one listener put it, "the human appeal of your voice."[36]

It was not only an American phenomenon. In Cuba, a highly popular program that showcased interviews with prominent political and cultural leaders based the discussion on questions listeners submitted through the mail. In Argentina, tango programs proved well suited to inspiring listeners to send letters, in part due to this musical style's links to Argentinian rural and folk culture to which so many listeners had ties. In fact, when Argentina's most popular radio station asked listeners to vote on their favorite new tango song, it claimed to have received more than 1.8 million mailed-in responses. In Mexico, programs that showcased traditional folk music inspired feelings of nostalgia in letter writers; one person celebrated how one such program had "spiritually transported" listeners back to small towns of their youth. From thousands of miles away, Pin Pin T'an apparently accomplished a similar feat of making her audience, spread out across much of Latin America, feel individually connected to her broadcasts.[37]

However, success at forging a sense of a personal connection with those letter-writing listeners is not necessarily indicative of whether the broadcasts were achieving their objectives. This verity holds even among the small cross-section of listeners who wrote celebratory letters. In the absence of visual cues, listeners are prone to drawing conclusions about the people they hear and information conveyed based primarily on the sound of the disembodied voice as opposed to the substance of the content. For example, one study showed that identical content spoken in different accents led listeners to draw different and contradictory conclusions about the intelligence of the speaker depending on what racial or ethnic group the accent signified in the mind's eye. Pin Pin T'an, who went by the Americanized name "Marjorie" (thus further masking the unseen radio host's ethnicity) may very well have been a "nice" and "gracious lady," as one listener from Vera Cruz, Mexico, wrote. Nonetheless, such presumptions and praise for her speaking style say little about whether the Basic English broadcasts were effectively teaching English. The fact that only a handful of letters praising the broadcasts were written in English, the vast majority instead having been written in Spanish, raises additional doubt about the initiative's overall effectiveness.[38]

The features of Latin America's linguistic, educational, and technological landscape further limited the radio initiative's ability to achieve its desired impact. The focus on Spanish-speaking Latin America, for example, cannot account for the extraordinary amount of linguistic diversity across the entirety of Latin America and the Caribbean. More than five hundred Indigenous languages were (and still are) spoken across the region, some of which are transnational. Speakers of Indigenous languages constituted demographic majorities in countries such as Bolivia and Guatemala, and in several provinces of other countries, such as Mexico. Marginalized in their respective nations, these Indigenous peoples lacked access to the economic, political, and educational opportunities available to the Spanish-speaking mainstream. Native Spanish speakers across Latin America were not necessarily any better positioned to engage the broadcasts. Class stratification and elitism limited access to the educational systems across the Americas. Only very small percentages of the respective national populations ever advanced beyond primary school. Those who did largely came from their countries' relatively small middle and upper classes. Among the poor who did enroll in school, the rate of absenteeism was frequently high. In Cuba, for example, between 1931 and 1953, only 35 to 45 percent of eligible children were enrolled in school, only about 13.5 percent of students who did enroll advanced beyond the fifth grade by the early 1950s, and less than 6.5 percent of Cuban children pursued a college education. Across Latin America, students who completed high school, to say nothing of completing college, were disproportionately from the upper classes. Aside from the limitations such statistics underscore regarding the Basic English project's potential to achieve its expansive objectives, that latter well-off and literate demographic, the most likely one to own a radio, was not one the Rockefeller Foundation typically targeted in the projects it funded. Even if, in the face of such obstacles, the poorer Spanish-speaking peoples across Latin America might have been interested in learning English through the Basic broadcasts, limited access to both an electrical infrastructure and shortwave radios outside of the largest cities further circumscribed the reach of those programs. Finally, in the case of Brazil, Portuguese rather than Spanish was the predominant national language. These combined linguistic, educational, and technological limitations deprived a significant number of students with the basis they needed to take advantage of—or even hear—the Basic English broadcasts.[39]

Even if situated in a more favorable societal and institutional context, the Basic system was still not necessarily well positioned to achieve the goal of becoming the "international auxiliary language" that its advocates envisioned. The foundation on which Ogden and Richards built Basic touted a deliberate circumscribing of the language's ability to convey conceptual meaning through the limited number of proscribed words. That foundation is inconsistent with fundamental processes of language-based communications. Basic, an artificial construct, lacks the dynamism and vitality that is inherent in any naturally occurring language. "Normal" language communication (in contrast to the artificiality of Basic) entails a variety of affective factors that go well beyond the rudimentary conveying of meaning. However, the deliberate limiting of Basic's vocabulary reflected the intent to eliminate the influence of feelings and distractions of the affective associations that accompany typical verbal communication. Consequently, Basic was not suited to convey the nuances of literary style or serve as a medium for expressing eloquence, wit, or intimate experiences. Such limitations made it a vehicle ill equipped to facilitate the very type of intercultural understanding and appreciation that stood at the core of the WWBF's mission. No amount of attention the WWBF gave to pedagogical approaches or reception theories could overcome this structural limitation built into Basic's very core.[40]

Moreover, the Basic system itself put forward a system of standardization based on a form of English closely associated with American and British spoken English (as opposed to other variants found in places such as India, Singapore, Australia, and elsewhere). The term "Basic" itself is an acronym for "*British American* Scientific International Commercial." Basic's fundamental structure and its very name exposed it to charges that the system primarily served the interests of American and British native English speakers and their respective nations' hegemonic aspirations. C.K. Ogden, in fact, openly admitted as much. For Basic to truly achieve the global potential he envisioned for it, Ogden advocated for the creation of a "strong 'Anglo-American Directorate' formed by the two governments to take control of its future expansion."[41]

In fact, well before the WWBF began its broadcasts, the English language's place as the national language of the United States rendered it a symbol of American self-interest and imperialist objectives in Latin America. Anti-Americanism in Latin America had, in fact, reached an apex during the 1930s, a culmination of recurring patterns dating back to the nineteenth century of blatant US violations of Latin American sovereignty, the racist and often brutal attitudes and actions that followed successive American occupations, and the crass materialism and economic self-interest that appeared to underpin so many of the offending American policies. As mid-century approached, a pan-Latin American vision of anti-Americanism manifested itself, which touted the superiority of Latin American values over US ones and advocated for multilayered resistance to US political, military, economic, and cultural aspirations. A pan-Latin American literary movement, rooted in social realism and popular among the same literate middle and upper classes that the Basic broadcasts hoped to reach, critiqued the harsh conditions and racist attitudes that were entangled with the overwhelming US economic presence. This demographic was the precise one that the Rockefeller Foundation's John Marshall accurately described as "pretty influential in Latin America." In this context, resisting the spread of English was another front on which many Latin Americans sought to fight US encroachments. As chapter 2 underscored, spoken English broadcasts heard over the radio in Mexico had the potential to provoke impassioned anti-American opposition and responses. While a handful of listeners from across Latin America wrote compelling letters offering the Basic broadcasts high praise, they cannot be considered representative of a Latin American listening audience. That audience, even if in a position to hear such broadcasts, could be just as prone to railing against the broadcasts if the programs were not ignored entirely.[42]

I.A. Richards, cultural blinders firmly in place, contemptuously dismissed any such linkages between efforts to spread Basic and anti-imperialist resistance. Whatever animosities the foreign policies and international activities of the United States and the United Kingdom created in various nations across the globe by the twentieth century were, Richards insisted, irrelevant. They did not render "Basic English" or the spread of the English language a corollary tool of "linguistic" or "cultural" imperialism. The "vociferous cries" of those making such claims came not from the peoples of the countries where Basic English was taught, but rather, Richards haughtily insisted, "from sensitive souls in New York or London." He had little patience for such warnings. "Let us be serious for a moment," Richards groaned. He belittled the claims by insisting the logic behind them was tantamount to claiming that reading Voltaire or Thomas Mann was the equivalent of "submitting to some sort of French or German conquest."

He proudly pointed to his work in China and resented being labeled some sort of imperialist for trying to improve education in that crisis-ridden country. If speaking English was so closely associated with Anglo-American imperialism, so presumably was cigarette smoking in China. It was a habit, he contended, brought to China by way of the "English speaking people's enterprise." And yet, Richards contended, impoverished Chinese who could barely afford to eat nonetheless purchased and smoked cigarettes rather than levying charges of imperialism against the Anglo-American-dominated tobacco industry in China. If the Chinese were so unmoved to protest the tobacco industry, surely the spread of English would not excite their animosities.[43]

That Richards said as much is an astonishing testament to his ignorance about the country in which he resided for many years. The Anglo-American-dominated cigarette industry in China had, in fact, become a quite visible and volatile focal point of labor unrest and anti-imperialist agitation during the first decades of the twentieth century. Richards's steadfast but ultimately ignorant defense of Basic thus lends credence to a later criticism levied against him by a former (and ultimately estranged) student, the renowned literary critic William Empson. Empson recalled Richards as preferring showmanship over substance. Richards was, Empson alleged, indifferent to argument and evidence if it meant being contradicted. Richards certainly proved that point when he favorably referenced Anglo-American domination of China's tobacco industry. If Basic's potential to proliferate across the globe depended on a keen understanding of and attention to the surrounding context, the co-creator of the Basic system only managed to betray his own alarmingly superficial understanding of relevant contexts on which any prospects for success ultimately rested.[44]

Despite the evident obstacles that limited the broadcasts' actual reach and potential effectiveness, the Rockefeller Foundation and WWBF's optimism for Basic endured into 1940. Both entities remained convinced that the spread of Basic English via radio promised to make a meaningful contribution to mutually beneficial cultural and economic exchanges in the service of fostering international peace and understanding. To be sure, the WWBF's efforts to learn and apply the lessons gleaned from research on listener reception and engagement, much of it Rockefeller-funded, were sincere. Nonetheless, culturally infused convictions about the combined power of Basic and broadcasting to eliminate international differences and reduce global frictions skewed their understandings. These beliefs undermined the genuine intentions to avoid simplistic and deterministic presumptions about the power of radio to singularly drive societal change. They ultimately focused on a relatively narrow slice of listeners that the Basic English broadcasts reached. In so doing, the principals in this project ended up looking past the more complex dynamics of the surrounding context that intersected with how listeners would—or would not—engage with the

Basic English programs. In this way, the Basic English broadcasts ultimately fell prey to the very overconfidence in the power of radio to singularly propel positive change those project principals had initially hoped to avoid.

* * *

By the time the United States entered the Second World War at the end of 1941, the WWBF's Basic English broadcasting initiative had thoroughly collapsed with little to show for its efforts. By the end of 1940, Lemmon had strained his relationship with both the Rockefeller Foundation and the Orthological Committee that had been established in Cambridge, Massachusetts, to oversee the development of the WWBF's Basic English broadcasts. He asked the foundation's John Marshall and committee member Charlotte Tyler (who also served as the WWBF's scriptwriter) to mislead the US Immigration and Naturalization Service about Pin Pin T'an's roles and responsibilities after it came to light that her hiring violated the terms of her student visa. The matter was ultimately resolved through proper legal channels, but the foundation provided no further funding to the WWBF's Basic English enterprise after 1940. T'an left the project altogether, first for a stint at the University of Chicago teaching Chinese to American soldiers and then to Swarthmore College where she taught English to Chinese naval officers.[45]

Her departure was of little consequence to Lemmon. By 1941, he had cast his lot with the more lucrative enterprise of using his WRUL station to help Britain's war effort against Germany. The US State Department, British Security Coordination, British Information Services, and the British Embassy's press office provided subsidies and personnel to WRUL for the development of foreign language programming directed at occupied Europe. British Information Services, in fact, began programming for WRUL and British Security Coordination prepared full scripts for broadcast. The station was ultimately placed under the authority of the US government's newly established Office of War Information for the duration of the war, which the United States entered at the end of 1941. WRUL's wartime activities inspired the US government to establish the Voice of America international radio network as its own radio outlet, with WRUL serving as the service's inaugural station before reverting back to Lemmon's control at the end of the war. Once back in charge, Lemmon dropped his aspirations for his station to serve as a radio university. He rebranded WRUL as "truly a weapon for peace and international understanding" and a vehicle that "batters away at the barriers of prejudice and international misunderstanding." English lessons vanished from the schedule. WRUL instead prioritized reaching international audiences in their own languages and began accepting commercial advertising in 1950. Lemmon ultimately sold the WWBF to the Metropolitan Broadcasting Corporation in 1960, the company that eventually provided the nucleus of stations that became the Fox television network in the mid-1980s.[46]

What became of the WWBF was a far cry from the vision Walter Lemmon shared with Woodrow Wilson at the end of World War I. At the start of the 1920s, Lemmon imagined radio's potential as a high-minded tool of education and cultural exchanges that shared with the world the best Americans could offer while bringing the peoples of the globe closer together in mutual understanding. By the 1940s, Lemmon transformed his radio enterprise into a weapon of the British-American war effort seeking to defeat Germany. This wartime and postwar fate of WRUL again underscores the way in which broadcasting is shaped by and beholden to the larger political, economic, and cultural forces that continually bear down on it. Lemmon's visions for the WWBF first articulated aboard the *USS George Washington* in 1919 and his efforts to execute that vision through the 1930s reflected larger aspirations for peace and understanding in an unstable world; WRUL's subsequent repositioning as a weapon in America's wartime arsenal reflected the changing dynamics of an utterly destabilized world that had once again gone to war. The specifics were adjusted to meet the demands of the moment.

Lemmon's two main convictions, however, remained consistent across the period: the righteousness of the American way and the WWBF's potential to spread those ideals globally. Yet, by choosing Basic English broadcasts as the starting point to achieve his global vision of mutual understanding and cultural exchange based on an embrace of American ideals, he selected a vehicle that was itself fraught with cultural and ideological baggage not necessarily complementary to his internationalist goals. Buttressed by convictions of the superiority of American and Anglo-American models of development and culture, the project was particularly ill-suited to engaging a vast audience in Latin America and the Caribbean during a period of heightened anti-Americanism in the region. Positive but ultimately unrepresentative listener response to those broadcasts highlighted what Lemmon and his collaborators wanted to believe about the broadcasts. The larger historical record leaves little evidence of these broadcasts having made much of a mark on their intended audience. The irony, of course, is that the expansion and growth of the Basic English broadcasts dedicated to fostering international peace and understanding paralleled the world's spiral into the Second World War. Along the way, the WWBF did not so much change the world around it for the better, but rather adapted to the changing US and global contexts in which it operated. In this way, the project's short-lived existence says far more about how Walter Lemmon and his cohort viewed radio and the world around them than it did about any educational value their Basic English radio initiative may have had.

PART III

Colonized Airwaves

CHAPTER 5

"A Workable Scheme to Quiet the Panaman Clamor"

US Radio Policy in Panama in the Shadow of the World Wars

By 1925, the frustration over radio's stalled development in Panama from Americans and Panamanians alike was palpable. The United States had used the leverage it derived from control over the Panama Canal and surrounding Canal Zone to stifle the growth of wireless telecommunications and broadcasting in the entirety of Panama. This throttling of radio in that small country came as the medium increasingly thrived in the United States, elsewhere in Latin America and the Caribbean, and throughout so much of the rest of the world. An editorial that appeared in a June 1925 issue of the English-language *Star and Herald*, published in the Canal's Pacific coast terminus Panama City, captured the sentiment. The only chance to hear programming was via a distant US-based station or another nearby country. The editorial, penned for an American audience, took pains to emphasize that Panamanians, like many of their Latin American counterparts, also had genuine interest in developing the medium. It speculated that radio receiver ownership would shoot upward if a broadcasting market was allowed to develop. Was Panama "to be forever isolated from the rest of the radio world"? Panama, after all, stood at the "geographical center" of the Western Hemisphere, situated on what was purportedly "the greatest commercial highway in world." But it might just as well have been in "the heart of darkest Africa" when it came to radio, the editorial grumbled.[1] Panamanians concurred. In fact, by the mid-1920s, Panamanian authorities increasingly threatened to ignore US-imposed radio policies if the situation did not change to allow Panama to control radio in its jurisdictions outside the

Canal Zone. That path of resistance, if followed, risked wreaking havoc on the isthmus's US-controlled radio system.

The sorry state of Panamanian radio by the mid-1920s was indicative of the challenges Panama faced in navigating its unequal relationship with the United States. Between 1900 and 1940, radio's emergence and evolution into a viable communications technology roughly paralleled Panama's establishment and development as a quasi-independent yet partly colonized nation. The United States—an emergent "Great Power" by the turn of the century—exerted an outsized influence over both transformations. In the early years of the century, the American diplomatic and military establishment saw little reason to be concerned with Panamanian perspectives and demands. Tiny and undeveloped Panama, they believed, would surely have remained a disgruntled province of Colombia were it not for the US machinations in 1903 that helped Panamanians achieve their independence. The price the United States insisted for that assistance was the right to build and control in perpetuity a transoceanic canal across the new country's midsection. As far as the Americans were concerned, Panamanians owed the United States deference and gratitude, not posturing and protests.

Over roughly the same period, radio outgrew its origins as an experimental and not wholly dependable alternative to wired telegraphy. By the start of the 1920s, it was a significant, reliable, and globally expansive communications medium. Here too Americans exercised an outsized influence over the development and proliferation of strategically and economically important wireless communications in the Western Hemisphere, which helped the United States solidify its ever-increasing economic and political dominance over so many of the nations in Latin America and the Caribbean. In this context, US officials—particularly those in the US Navy—saw in a prostrate and partially colonized Panama a strategically ideal locale to exert overarching control over the developing medium that complemented larger hemispheric interests and aspirations.

US radio policy, in turn, became a focal point of nationalist and anti-American resistance that increasingly defined Panamanian politics between the two world wars. From Panama's perspective, American radio policy symbolized yet another example of repeated affronts to its sovereignty and independence. Especially egregious in Panamanian eyes was the US determination to deny Panama authority over radio in the entirety of the country. Panamanian officials invariably rejected American claims that the provisions of their 1903 treaty, which granted the US authority to build and operate the canal, also legitimated the Americans exercising country-wide control over radio. That US claim underscored greater Panama's subordination as an informal American colony adjacent to the actual colony of the Canal Zone. One contemporary foreign policy expert compared the overall US position in Panama to the ways in which the British and Japanese exercised indirect rule over Egypt and Manchuria, respectively.[2] Radio policy was one example that illustrated how the United States exerted control beyond the confines of the Canal Zone.

Panama's relative weakness and subordination vis-à-vis the United States notwithstanding, Panamanian authorities had options. Panamanian officials had no choice but to confront increasing popular frustration over the circumscribing of sovereign rights enjoyed by other independent nations. Panamanian leaders deemed too deferential to American strong-arming over radio risked domestic political blowback. Radio, however, was a relatively simple and accessible technology that was also vulnerable to interference if multiple transmissions crossed the same frequency. Through either the threat of interference or the deliberate infliction of it, Panama thereby had the means to effectively resist American radio mandates. American diplomats in turn responded, albeit belatedly, to Panamanian disaffection and resistance. They did so by trying to adjust US radio policies to account for Panamanian perspectives and interests. This pivot in US policy captured a broader evolution in US-Latin American relations that began in the 1920s and culminated in the 1930s with the "Good Neighbor" approach to US foreign policy in Latin America and the Caribbean.

However, inconsistencies and incoherencies in the application of Good Neighbor approaches to US radio policy in Panama often worked at cross-purposes with achieving the positive outcomes this policy pivot imagined. Most notably, the State Department's shift to a more accommodating diplomatic stance triggered a rupture with its military counterparts, particularly the Navy. Having lost its influence at the negotiating table in the face of the Good Neighbor policy's ascendance, the Navy shifted its fight to retain control over radio in Panama to the arena of US domestic politics. In so doing, the Navy sabotaged a series of 1936 agreements intended to find mutually acceptable solutions to long-standing radio and other grievances. This US-Panama radio dispute ultimately endured into and beyond World War II. Conflicts over radio consistently enflamed deeply rooted disagreements over policy priorities and larger worldviews. The resulting fissures did not just pit Americans and Panamanians against one another, but propelled divisions within the American and Panamanian camps as well.

"Complete and Permanent Control": The Origins of US Radio Policy in Panama

The fifteenth day of August 1914 should have been a celebratory day for the United States. The just-completed Panama Canal opened to shipping traffic for the first time. Completing this massive project involved surmounting an array of daunting engineering and tropical disease challenges that felled an earlier French-led effort. During the preceding decade, the United States overcame those formidable obstacles and completed the construction of the spectacular waterway. In 1897, renowned naval theorist Alfred Mahan excitedly anticipated the importance of a future US-controlled canal across Central America. He proclaimed the eventual completion of such a passage would advance "by thousands of miles the frontiers of European civilization in general and of the United States in particular, that it knits together the whole system of American states enjoying that civilization as in no other way can they be bound." Ira Bennett, a *Washington Post* journalist who in 1912 was given privileged access to government records to write a book on the Canal's construction, effused that the waterway demonstrated how "the practical genius of the American industrial world rose to meet the new and extraordinary conditions under which the work progressed to a triumphant conclusion." In January 1914, with the Canal only months away from opening, southern industrialist R.H. Edmonds published a piece in the trade magazine *Manufacturers Record* in which he celebrated the Panama Canal's "limitless possibilities for . . . human advancement," "the advancement of civilization," and the "advancement of human progress." To Edmonds, who echoed the rhetoric of many of his contemporaries about radio's significance, the Canal's opening promised to "break down barriers" and "crystallize the thoughts of the people of every land," spanning "every hamlet, from the far Himalayan Mountains to the wilds of the Andes, from Siberia to the utmost limits of South America."[3]

Despite the years of anticipation, the Panama Canal's opening on that fifteenth day of August was far from the biggest news of the moment. Most attention was instead directed toward an ever-expanding European war that erupted during the last week of July 1914. The June 28 assassination of Austria-Hungary's crown prince at the hands of a Serbian nationalist triggered military mobilizations across Europe that ultimately resulted in a spate of war declarations. The war news dominating the front pages in the middle of August underscored Austria-allied Germany's audacious invasion of neutral Belgium to advance into France. "Germans Harried, March Ahead" blared the headline splashed across the top of that Saturday's *Chicago Daily Tribune*. A similar tone was struck by newspapers across the country. During the preceding weeks, nations had rushed to join their allies on one side or the other. By mid-August, Russia, France, and England (the so-called Entente) were pitted against Austria-Hungary and Germany (the "Central Powers"), with an array of other smaller powers (including Serbia, of course) joining the fray as well. The conflict continued to widen into a world war and before month's end Britain's Asian ally Japan had aligned with the Entente. News of the Canal's opening during those middle days of August was typically buried several pages further into a daily paper.[4]

The outbreak and widening of war in the summer of 1914 did not, in fact, have any meaningful bearing on US policy preferences for radio in Panama. American radio anxieties had been increasing since at least 1910. That year, the Panamanian government gave United Fruit, a US-based multinational company headquartered in Boston, permission to build a station in Colón. It did so, however, in the absence of any consultation with US officials. It was not the first time United Fruit and Panamanian authorities struck a radio agreement among themselves. Since 1905, United Fruit had operated a station on its holdings at Bocas del Toro; it was one of the first commercial stations in Latin America. The concession had been granted before the larger strategic significance of the medium was fully appreciated and before the United States conceived of a radio policy for Panama that accounted for that significance. As appreciation for the strategic significance of radio increased in the subsequent years, the United States tolerated the Bocas del Toro station's continued operations; it had not interfered with the development of a larger Navy-controlled system of communications in and around the Canal. However, US officials were not so sanguine in 1910 when United Fruit received permission from Panama to build a new station in what would be the Canal's Atlantic coast terminus city of Colón. Panama's continued unilateral authorization of new stations, they worried, risked inflicting debilitating interference over the Canal Zone's emergent wireless communications network. At that point, US officials concluded that in a country as small as Panama with the US-governed Canal Zone bisecting it, singular American control over radio was required. The medium's border-crossing propensities and vulnerability to static interference when too many transmissions crisscrossed the finite number of usable frequencies demanded as much. Once war erupted in 1914, the inclination of belligerent actors to deliberately disrupt communications to gain an advantage during wartime merely added another bullet point to an ongoing argument that the United States was already determined to win.[5]

Such strong US opposition to United Fruit's radio expansion in Panama was the exception for Latin America. For its part, the US government typically encouraged American companies such as United Fruit to strike agreements with other regimes throughout the hemisphere to build radio networks. In this regard, though not an official entity, United Fruit was an example of a so-called chosen instrument, a nongovernmental entity that US officials often tapped to help achieve larger foreign policy goals. In this instance, the goal was to keep other foreign commercial and military competitors out of Latin America and thereby ensure continued US dominance in the hemisphere. Outside of

Panama, such support was typically consistent with the Navy's vision of a so-called Monroe Doctrine for radio. For United Fruit's part, the Boston-based company's extensive landholdings and massive plantations that grew bananas and other tropical fruits for export into the US market benefited from the ever-more rapid and reliable communications that radio provided. Such links helped reduce costs and increase profits by, for example, ensuring dock workers were positioned to unload perishable cargo as soon as a ship entered a port or, alternatively, not paid to stand on the dock doing nothing if a ship was significantly delayed. For these reasons, United Fruit established the first radio stations in Central America and was the first to provide a commercial radio service connecting the United States to the region. For a company like United Fruit, radio was far superior to wired telegraphy, with the latter's physical infrastructure vulnerable to repeated incapacitation in extreme tropical environments prone to dramatic weather events like hurricanes and flooding, boring insects, and the steady corrosion of wires from the humidity, not to mention the possibility of deliberate sabotage.[6]

It was those same characteristics that made radio an appealing medium to the Panamanian government. The fledgling regime saw in radio a tool to solidify and extend its authority over the more distant reaches of the newly independent and largely undeveloped country. Panama had only achieved independence in 1903, with American assistance that was predicated on getting permission to build and control a canal. The United States then forced the heretofore rebellious Colombian province to sign the 1903 Hay–Bunau-Varilla Treaty affirming that permission as its price. The previous year Colombia had rejected a similar accord to build a canal across what was then its northernmost province. The Panama treaty was comparable to its Colombian predecessor, with two notable exceptions: a limited ninety-nine-year lease to the Canal and surrounding Canal Zone stipulated by the proposed accord with Colombia was revised in the Panama version to allow for US control in perpetuity, while the United States also pledged itself to protect Panamanian independence.[7]

That origin story and successful construction of the Canal gave the United States a strong sense of entitlement over Panama. By the turn of the century, US policymakers typically held strong convictions regarding the United States' professed rights to influence and intervene elsewhere in Latin America and the Caribbean. Given the central role the United States played in helping secure Panamanian independence and building the Canal, such sentiments were even stronger toward Panama. Radio, in turn, became a focal point in exercising this US power over the semi-colonized country. In 1911, the US Congress had appropriated $1 million (more than $32 million in 2023 dollars) to fund the construction of Navy-operated high-powered stations connecting the Panama

Canal Zone, Hawaii, and the Philippines. Late that year, Panama subsequently announced its intention to build its own government-run station, again in the absence of consultation with the US authorities. This move thereby antagonized already escalating American concerns that Panamanian radio plans threatened the functionality of the easternmost link in the Philippine-Hawaii-Panama network. The Navy responded by relaying its "urgent" concerns to the State Department. By the autumn of 1913, Secretary of the Navy Josephus Daniels bemoaned the "state of radio lawlessness outside the limits of the Canal Zone," the "unnecessary number of stations in the same locality," and the "great annoyance" they posed. A US government "monopoly" over Panamanian radio, Daniels argued to Secretary of State William Jennings Bryan, would be "the surest way of obviating that condition."[8]

The Americans had no interest in Panama's assurances of cooperation over radio. The 1903 Hay–Bunau-Varilla Treaty, according to the Navy, gave Panama no say in the matter. It pointed to the clauses that allowed the United States to act "as if it were sovereign" beyond the Canal Zone when it came to the "use, occupation, and control of any other lands and waters outside the zone . . . which may be necessary and convenient for the construction, maintenance, operation, sanitation and protection of the said Canal." Citing these clauses to assert control over radio captured that emergent view that invisible radio frequencies were themselves a natural resource akin to and intertwined with the water and land over which transmissions traversed. The Navy further stipulated that when the United States made such a claim, it resulted in Panama's "entire exclusion" from being able to "exercise . . . any such sovereign rights, power, or authority" in the matter. An exhaustive analysis by the State Department's Office of the Solicitor in 1912 affirmed, at least in its view, the legal merits of the Navy's position on radio in Panama. The State Department then tasked the Canal Zone's Joint Army and Navy Board to draft a proposal for Panama's consideration that reflected this conclusion. When the Joint Board completed its proposal in the autumn of 1913 amidst Daniels's complaints of "radio lawlessness," it unsurprisingly called for a complete US monopoly over radio in Panama.[9]

Panama predictably balked at this demand. Panamanian complaints about American abuses of Panama's sovereignty under the guise of the 1903 treaty were hardly new and had been levied against US policies toward ports, custom houses, tariffs, and post offices for as long as Panama had been recognized as an independent nation. But at least those earlier grievances typically singled out US policies within the Canal Zone. For radio, the United States now demanded authority that extended beyond that zone into broader Panama. In fact, the National Liberal Party administration of President Belisario Porras Barahona ascended to power in 1912 on a platform that prioritized standing

Figure 10. Belisario Porras Barahona. Courtesy of the National Photo Company Collection, Library of Congress Prints and Photograph Division.

strong against further American encroachments on Panamanian sovereignty and independence. This platform earned the Liberals a broad base of working-class support. Porras himself rose to national prominence as an anti-Colombian independence fighter. In that vein, he was also a long-standing and outspoken critic of the Hay–Bunau-Varilla Treaty and became the opposition leader to the ruling Conservatives before winning the presidency himself.[10]

Porras, to be clear, supported an American presence in Panama. He acknowledged that his country's independence owed much to the United States and its ambitions to build a canal. "Panama exists by and for the Canal," Porras conceded. In fact, the Belgian-educated Porras shared with many of his fellow internationally schooled Panamanian ruling elite, Liberal and Conservative alike, the belief that Panama absolutely required American support and resources to modernize the nation. At the same time, Porras had to balance this desire for US support without wholly alienating influential domestic constituencies that recoiled at the reality of US domination over the still-new nation saddled with such unusual limitations on its independence. This convergence of pressures drove, in the words of one scholar, Porras's "zigzagging posture" toward the

United States in determining when to concede to American wishes, when to stand strong, and—most often—when to accept a deal that landed somewhere in between.[11]

For the Porras administration, navigating American radio demands became one such exercise in zigzagging. The day before the Canal opened, Porras approved a decree recognizing temporary US authority over radio that would stand only for the duration of the emergent world war. The United States rejected that order, insisted it not be published, and asserted it would accept nothing less than a permanent ceding of control. The Porras administration relented and signaled it would be willing to entertain such a decree, but for a price. In exchange for such an unqualified permanent ceding of authority, the American-educated Secretary of Foreign Affairs Ernesto Lefevre de la Ossa secured an American commitment to build three radio stations far from the Canal Zone. Lefevre, a longtime Porras ally, framed the demand for those stations as complementary to his government's broader state-building aspirations. Lefevre argued that such stations would provide fast and reliable communications with the more remote regions of Panama heavily populated by Indigenous peoples. In so doing, the requested stations promised to protect both troops and inhabitants from an "Indian massacre," to "civilize the Indians," and to "open these fine lands to civilization."[12]

Lefevre's deliberate use of this language highlights a resistance strategy that the colonized at times deploy against their colonizers. Evident complicity of native authorities with the colonizing power, in Panama and elsewhere, was not and never was reflective of total submission. In this instance, Lefevre deployed language that echoed the ideological justifications of dominance over an inferior "other" in the pursuit of beneficial technological progress. It was language that the Americans themselves long used to justify their own imperial projects, including the one in Panama. As such, this language was bound to resonate with the intended audience of US officials. In so doing, Lefevre's gambit was exemplary of how the colonized can choose to selectively imitate the colonizer to minimize unwelcome interference and wrest a degree of control away from the colonizer. Moreover, as Lefevre's objectives also make evident, the purpose of such imitation was not to make a claim to be the same as the colonizer, but rather to emphasize and assert a unique identity and interests separate from the colonial authority. This imitation most certainly came easy to Lefevre, given that Panama, like the United States, viewed race through a prism of a hierarchy that separated superior "civilized" peoples from inferior "uncivilized" and backward peoples. It was a synergy of views that, in fact, became another enduring source of Panamanian animosity toward the United States amidst policies that deliberately discriminated against supposedly inferior Panamanians.[13]

Ultimately, the Americans acquiesced to Lefevre's request to build stations for the Panamanian government and Panama then ceded full authority over radio to the United States. Secretary of State William Jennings Bryan grumbled a bit, to be sure. He was aghast at Panama's inclination to "bargain and negotiate about such an important matter." In Bryan's view, US control over radio was essential not just to the safety of the Canal but to "the guaranteed independence of Panama." The irony of the US position—insisting that the United States could only guarantee Panama's independence if Panama agreed to circumscribe its independence—was evidently lost on the secretary of state. Nonetheless, on August 29, 1914, Porras signed a revised radio decree that capitulated to US demands. That said, a decree proclaimed solely under the auspices of the president's singular authority could presumably be withdrawn through the exercise of that same singular authority. For the present moment, however, Porras had suitably mollified the Americans.[14]

Alas, the sweeping 1914 decree ultimately did not provide the desired long-term elixir to quell the US-Panama radio conflict. Instead, the first two stations the United States agreed to build in exchange for that decree provided the fodder for resurgent animosities soon after the war ended. Those stations were built at La Palma and Puerto Obaldía, as per Panama's request. Since the stations were to serve Panamanian interests and objectives, Panama had paid for the acquisition of the land on which those two stations were built (the United States already controlled the property on which it constructed the third station at Punto Mala). Those stations, which the Americans would operate on behalf of Panama, were on track for July 1919 and July 1920 openings, respectively. La Palma lay more than a hundred miles southeast from the Canal's western terminus on its Pacific side; it was only about fifty miles from the nearest point of the Colombian border. Puerto Obaldía was situated on Panama's Caribbean coast and less than two miles from the Colombian border. The nearest point of the Canal Zone lay some 150 miles away and the Canal's Atlantic terminus was 170 miles away. However, the Navy quietly pushed the United States to claim actual sovereignty over the territory on which those two stations stood (approximately fifteen acres for each station), despite the fact that Panama had purchased the land. In December 1918, State Department counsel concurred that such a claim would be legitimate under the terms of the 1903 treaty. At the start of 1919, Under Secretary of State Frank Polk, then serving as acting secretary of state, confirmed to Navy secretary Josephus Daniels the State Department's overall support of that position.[15]

A stunned Panama did not learn of the Navy's sovereignty gambit until early 1920. With the Entente defeating the Central Powers by the end of 1918, the wartime emergency could no longer provide even a shroud of justification for the US actions. When informing Panama of the US intent, the American Minister in Panama William Jennings Price acknowledged that the Panamanian government purchased the land, but nonetheless claimed that the 1903 treaty and its broad stipulations allowed for the sovereignty claim. Outwardly, Panama's Foreign Ministry chose to tread lightly in its response. Rather than have Lefevre respond directly, his assistant offered a tepid reply that side-stepped the sovereignty issue by acknowledging the United States would operate the station while also reminding Price that there was no agreement between the two nations confirming the US interpretation of its treaty rights. That apathetic reaction from a lower-ranking official, Price concluded, was "as satisfactory a response we could hope for."[16]

Panama was, in fact, far from satisfied with the American machinations. The timing of the sovereignty claim was especially provocative. President Porras was nearing the close of his second term at end of 1919 and angling for an unprecedented third one in the forthcoming 1920 elections. That he occupied the office of the presidency in 1919 was unexpected. He had been constitutionally prohibited from seeking reelection to a second consecutive term in 1916. Porras's hand-picked successor won the 1916 contest, but unexpectedly died two years later. The sudden presidential vacancy enabled Porras to deploy some dubious political maneuverings to reclaim his former office. By the start of 1920, now eying a third term as the end of his truncated second term approached, the president concocted a plan to skirt the constitutional restriction against reelection: he resigned the presidency in January, handing the office to his ally Ernesto Lefevre, and announced his candidacy for president in the upcoming 1920 elections scheduled for the summer. Into these political maneuverings the United States dropped its explosive claim of sovereignty over Panamanian lands far from the Canal. It was not a good look for a leader who established his reputation as a committed defender of Panamanian sovereignty against American encroachments. The US-inflicted humiliations were, in fact, piling up by 1920. In addition to the 1914 decree issued under Porras's signature, the forced disarmament of Panamanian police in Colón and Panama City in 1916 and the American occupation of Chiriquí Province in 1918, both in response to allegations of anti-American threats and violence, also occurred under Porras's watch. The president's impotence in the face of the American-imposed police disarmament ultimately drove his ruling National Liberal Party to split into two rival factions, one of which opposed Porras. Porras ultimately prevailed at the polls in 1920, albeit with some underhanded manipulation of the electoral process to ensure his victory.[17]

By 1920 it was impossible for Porras to avoid the fallout from yet another humiliation at the hands of the Americans. The anger in Panama's National

Assembly was especially intense. Just days after Porras's third term started on October 1, 1920, Representative Hector Valdes, a National Liberal like Porras but from the rival faction of the party, introduced a resolution that, if passed, would effectively censure the executive branch over its repeated yielding to American demands on issues of sovereignty. "Sovereignty does not depend on the American government," Valdes fumed. He lambasted the US Navy's insistence on "taking the lands of the Republic" as "a stupid abuse." Valdes acknowledged, given the decidedly unequal US-Panama relationship, that Porras alone was not to blame. The legislator conceded that the president and his predecessors had good intentions in the context of severely limited options and power. Still, Valdes argued, officials throughout the executive branch, including Porras, had been overly accommodating toward unreasonable American demands and bore some responsibility for the United States' present overbearing posture. "There are no semi-sovereign nations," Valdes exclaimed. "PATRIOTISM," Valdes continued (with that capitalization used in the transcript of his remarks), demanded Panama's leaders do more to protect their country's sovereign rights.[18]

Ernesto Lefevre—now the former secretary of foreign affairs, the outgoing interim president, and one of the targets of Valdes's tirade—followed with his rebuttal. He assured the legislators that he understood and shared their frustrations. Lefevre, however, vigorously defended the records of Porras and his predecessors (of which he was now one). President Porras had never yielded easily to US demands, Lefevre reminded the legislators. He had ceaselessly worked to convince the United States to agree to arbitration as the means of reconciling divergent interpretations over treaty rights. Heatedly defending his own patriotic bona fides as a former president, secretary of foreign affairs, and before that minister of posts and telegraphs, Lefevre was adamant that he too "always zealously defended Panamanian interests, despite being a great admirer of the United States." Panama's "original sin," Lefevre argued, lay not with vacillating leadership but with the unavoidable signing of the "humiliating" and "embarrassing" 1903 treaty that Porras himself opposed. Had Colombia not been poised to retake the isthmus, Lefevre recounted, Panama would never have agreed to a treaty containing "so many clauses contrary to national interests." Ultimately, Lefevre concluded, Panama had few meaningful options when confronted by an uncompromising United States that wielded broad treaty rights and a preponderance of power. Beyond lodging repeated protests and demands for compensation, Panama could only wait for "America's spirit of justice to prevail over all else."[19]

Though that wait promised to be long, Lefevre's rebuttal won the day. Lefevre reportedly became quite animated as he reached the end of his address. Upon ending his remarks, Lefevre composed himself and was greeted with a round of applause as he returned to his seat. The Spanish language section of Panama City's *Star and Herald* described Lefevre's speech as the "great topic of the day," praised the "naked" and "cold blood[ed]" expression of his opinions. Lefevre, described as "the one who knew the most about the matter," even managed to reunify the factionalized Liberals. In so doing, the outgoing acting president staved off the censure resolution in what turned out to be his final official act. He left government after his presidential stint and died following surgery two years later at forty-six years of age. Lefevre's impassioned final speech notwithstanding, Panama was not content to wait for any "spirit of justice" to wash over the Americans and change their views of treaty rights. On January 31, 1921, Panama informed the United States that it unequivocally rejected the American claims to sovereignty over the stations at La Palma and Puerto Olbadía.[20]

The Navy remained indifferent. Instead, it brought the dispute to a head at Puerto Obaldía in August 1922. By this point, Panamanian anger at the United States had moved far beyond the sovereignty claim alone. American authorities closed access to Puerto Viejo, a small bay proximate to the station that locals had used for generations to dock their boats. Adding injury to insult, American troops had haphazardly fired rifles to chase cattle from what had heretofore been public grazing lands, with nearby houses allegedly suffering damage from the stray bullets that hit them. Then, amidst this simmering Panamanian resentment, a Colombian thief had crossed the nearby border and stole from residents of the area. With Panamanian police in pursuit, the fugitive took refuge on the station's grounds. US naval authorities, however, denied entry to the pursuing Panamanian authorities on the basis that they lacked jurisdiction on territory over which the United States was sovereign. Secretary of Foreign Affairs Narciso Garay Diaz was aghast at the Americans' "erroneous" position and "absolute disregard for the authorities which have been lawfully constituted by the sovereign power, which is the Republic of Panama." American Minister John Glover South, who had since succeeded William Jennings Price, largely confirmed the accuracy of Garay's description of the US actions but dismissed the substance of the protest on the basis that US claims were legitimate and not in any way erroneous. Panama, as far as South was concerned, had no choice but to accept the US position.[21]

By this time, broadcasting was also emerging as a contentious issue in US-Panamanian relations. World War I coincided with the appearance of what became the hallmark practices of radio broadcasting, such as playing phonograph records over the airwaves for entertainment as the young Walter Lemmon, discussed in chapter 4, did during his 1919 journey to Versailles with Woodrow Wilson aboard the *USS George Washington*. After the war's conclusion in 1919, those practices spread widely throughout the world, including across Latin

Figure 11. John Glover South. Courtesy of the George Grantham Bain Collection, Library of Congress Prints and Photographs Division.

America. The suffocating American radio policy in Panama made that country a notable exception. Even Americans were not shown any favor. S.E. Farwell, president of the Boston-Panama Company, learned as much when in August 1922 he futilely queried US authorities about building a station for his fruit and produce company just as United Fruit had done earlier in the century. Another request came in November 1922 from Panama's first amateur radio organization, comprised of both American and Panamanian membership. The group observed that practices regarding transmission power, wavelength allocations, and station location had facilitated the recent development and transformation of broadcasting throughout the world in a way that allowed many stations to operate simultaneously in a single area without interfering with one another, thereby enabling amateurs to communicate with each other across the globe. "Why not PANAMA?" they asked.[22]

Indeed, why not Panama, Panama's Minister in Washington Ricardo Joaquin Alfaro Jované also questioned at the end of 1922. Already primed for a fight over the US push for sovereignty over Puerto Obaldía and La Palma, Alfaro fumed at the American authorities' unilateral rejection of the Panama Radio Club's request to allow amateur activities. The minister saw the US obstruction as indicative of a deliberate effort to stifle the "great development" of broadcasting across the entire country. Alfaro marveled that in the United States broadcasting entailed "concerts, speeches, conference, reports, notices, and everything that may bring knowledge or amusement to the citizens of every class, condition and occupation." Anyone in the United States, citizen and noncitizen alike, was free to take advantage of broadcasting, and yet US authorities blocked Panama's very first amateur radio club from doing what amateurs across the United States and the rest of the world were already doing. The United States was deliberately preventing Panama "from enjoying benefits of modern science." How was it, Alfaro questioned, that the types of broadcasting freely permitted in the United States "cannot be done in its own country by the Government of the Republic of Panama"? The time had come, Alfaro insisted, to cancel the 1914 decree, which after all had been issued under the duress of US pressure amidst a now-past wartime emergency. That outdated decree, he continued, should be replaced with a mutually satisfactory radio agreement that accounted for Panamanian interests. If the Americans refused to change course, the minister threatened, Panama would rescind the decree and unilaterally pursue its own radio policies.[23]

The US Navy and Canal Zone authorities again dismissed the grievances. Broadcasting, they argued, was an unnecessary luxury for Panama. The American authorities repeated a familiar refrain: too many stations in too small a territory created too great a risk of causing debilitating interference with critical Canal communications. For good measure, Canal Zone governor Jay Morrow added that the unfavorable tropical climatic conditions quickly corroded receivers in a country that supposedly lacked native talent and a mass audience to support the medium. Instead, Morrow proposed permitting receiver ownership and allowing a limited broadcasting service, which the Navy subsequently agreed to provide. The service, launched in early 1923, promised programs that would be, as Secretary of the Navy Edwin Denby put it, "suitable to the general wishes of prospective listeners, and which may be furnished by the Panamanians themselves." Morrow figured listeners would quickly realize that stations heard from the United States, Cuba, or elsewhere in the proximate region better satiated their interests. The Navy broadcasts were, in Morrow's view, nothing more than "a workable scheme to quiet the Panaman [*sic*] clamor for alleged benefits of a broadcasting service." It would suffice "until such time as will prove to them there is no field for commercial broadcasting in their area."[24]

The clamor did not subside. In fact, by November 1923, Panama was the fifth largest export market for American receivers, even though the only stations one could hear other than the Navy outlet lay outside of Panama. Although stopping short of canceling the decree, Panama had moved evermore defiantly throughout 1923. Early in the year, US officials expressed irritation that the Panamanian government, again without consultation, ordered receiving sets, one of which would be installed in the presidential palace. In spring 1923, Panama flatly rejected an ill-advised US Navy-inspired proposal to replace the 1914 decree with a more detailed enumeration of rules and regulations that confirmed the United States as the sole radio authority in Panama. For good measure, in its rejection Panama forcefully emphasized its fundamental disagreement with the US claim that the 1903 treaty provided a legitimate basis for the United States to claim expansive authority over radio. The only basis for such control, Panamanian authorities asserted, was the 1914 decree. Panama then once again threatened to revoke it using the same presidential authority under which it was issued unless the Americans negotiated in good faith toward a new bilateral radio agreement. Then, in the summer of 1923, Panama announced its intent to unilaterally authorize the installation of wireless equipment on Panamanian vessels. Panama also declared it would not enforce proposed US radio regulations in Panamanian jurisdictions until the United States better accommodated Panamanian concerns.[25]

Tiny Panama had big leverage. When it came to radio, disregard of the US-imposed rules could itself be a form of effective resistance. Panamanian disregard for US radio regulations underscored the medium's vulnerability to the very interference that the Navy hoped to avoid. To this point, though, American officials in Panama conceded that Panama's contravention of US radio rules had not yet disrupted US-controlled radio communications networks in Panama. Nonetheless, the interference-free status quo could not be expected to last much longer if Panama continued to act on its own and without coordination. In summer 1923, Secretary of State Charles Evans Hughes relented in the face of Panamanian pressures. Hughes informed the Secretary of the Navy Edwin Denby that even though he and the State Department agreed that the United States was justified to assert its authority over radio in the entirety of Panama, he was nonetheless authorizing negotiations with Panama. The objective was to agree on a new treaty intended to resolve that long-standing dispute. Nine years of heavy-handed US radio policies designed to secure permanent and absolute control over the medium in Panama had accomplished little more than fueling and intensifying the broader anti-American nationalism that suffused Panamanian politics by the early 1920s.[26]

"Trust the State Department": State-Navy Rivalry and US Radio Policy toward Panama through 1933

The Navy balked at the prospect of negotiations with Panama over radio. Instead, it pushed for a presidential executive order to invoke the terms of the 1903 treaty and impose complete US authority over radio on Panama. In strictly legal terms, the State Department concurred that, in the absence of Panamanian consent, a presidential executive order so crafted was a legitimate instrument under international law to impose US radio policy and its related territorial sovereignty claims. But was it wise? In 1922, Dana Gardner Munro, a young analyst in his early thirties who was new to the State Department's Division of Latin American Affairs, doubted it. Munro was among the first of many officials to question the Navy's preferred and unrelenting hardline approach to radio in Panama that had produced few dividends to that point. Though not doubting the legality of the executive order the Navy desired, he counseled his boss, the thirty-nine-year-old newly appointed Assistant Secretary of State Leland Harrison, to avoid going that route "unless it is really necessary." Such a unilateral action would do nothing to solve the problem and more likely aggravate it by provoking "a strong protest and make our relations with Panama even more strained than they are at present."[27]

Munro's counsel reflected the priority he placed on understanding and accounting for Latin American perspectives in the pursuit of US policy goals. Prior to joining the State Department in 1919, Munro had spent two years living in Central America, learning Spanish and researching Latin American political and economic development on the way toward earning his economics doctorate from the University of Pennsylvania in 1917. Before being assigned to the Division of Latin American Affairs in 1921, he worked as an economist in the State Department's Foreign Trade Office focused on Mexico and the Caribbean

Figure 12. Dana Gardner Munro. Courtesy of Harris and Ewing Photographs, Library of Congress Prints and Photographs Division.

region. Munro then accepted a posting at Valparaiso, Chile, where he served as an economic counsel. He was inclined to disparage "tactless" US officials who paid no heed to the interests and perspectives of their Latin American counterparts. Eventually leaving government in 1932 to embark on what became a successful career in academia as a prominent scholar of US-Latin American relations, Munro continued to emphasize the importance of being able to "imagine what went on at the other end of the cable line." His counsel to Assistant Secretary of State Leland Harrison regarding radio in Panama did just that.[28]

Dana Gardner Munro's approach to diplomacy captured the broader albeit inconsistent efforts of a new generation of policymakers to change the tenor of US-Latin American relations during the 1920s. As anti-Americanism intensified across Latin America and the Caribbean, pressure increased to pursue more noninterventionist and cooperative foreign policies toward the nations of the region. It was, the rationale went, a more effective way to protect and advance American economic and strategic interests. Historians such as Alexander DeConde, Alan McPherson, and Bryce Woods are favorably inclined to give some credit to the Republican administrations of that era for the gradual development of the nascent philosophies, approaches, and policies that ultimately became formalized in 1930s through Democratic president Franklin Roosevelt's "Good Neighbor" approach to US-Latin American relations. The Good Neighbor precedents attributed to Roosevelt's Republican predecessors included compensating Colombia $25 million in 1921 for the US role in severing Panama from its grasp and building the Canal. The withdrawal of Marines from the Dominican Republic in 1924 and the Republican-negotiated agreements that led to the withdrawal of Marines from Nicaragua and Haiti in 1933 and 1934, respectively, are also cited in a similar vein. Munro himself helped negotiate the terms of the withdrawal from Haiti during his final government posting as the American Minister in Haiti. Munro, later as an academic in his own right, was among the first scholars to underscore the 1920s origins of the Good Neighbor policy in his scholarly writings (in which he acknowledged his own "minor role in policy formulation" toward Central America and the Caribbean). Historian Irwin Gellman, however, is adamantly opposed to awarding such credit to that decade's Republican administrations given the overall incoherence and inconsistency of their policies and approaches. Indeed, in the case of the troop withdrawals from the Dominican Republic and Haiti, the Marines' presence in those countries reflected earlier occupations authorized by Republican presidents Warren Harding and Calvin Coolidge, who were in office between 1921 and 1929.[29]

US radio policy toward Panama captured these inconsistencies and contradictions. For example, Francis White, Munro's superior and one of the most influential State Department specialists in Latin American Affairs, disdained Canal Zone governor Jay Morrow for viewing the Panamanians as "an inferior race who should be kicked about and told what to do." At the same time, White also refused to disavow the right to unilaterally intervene in Latin American nations if the United States perceived a threat. Allen Dawson, who like Munro and White served a stint as the chief of the Latin American Affairs Division in the 1920s, believed Panama's 1914 presidential decree "went beyond our needs" and "retarded the development of commercial radio in Panama unreasonably." But the more accommodating attitude that Dawson embraced from Washington was not necessarily reflected in the actions of the American Minister John Glover South. In 1927, South went so far as to lecture Panama's president Rodolfo Chiari Robles for "unwittingly" undermining US radio interests in Panama and warn him that Panamanian authorities were "not in a position to judge . . . these interests" because when it came to radio policy "only the American Government [was] qualified to do so."[30]

The ill-fated Alfaro-Kellogg Treaty of 1926 is a testament to the incoherencies that also afflicted US radio policy toward Panama. This accord was the outcome of the negotiations that Secretary of State Charles Hughes had authorized in 1923. The final draft went far beyond radio; it attempted to address a wider range of Panamanian grievances and conflicting interpretations of treaty rights. The accord utterly failed to do so. Panama's National Assembly found the treaty so offensive that it refused to even put it to a vote. This rebuke was not surprising. The treaty's clauses addressing radio gave lip service to greater Panamanian control, but explicitly left the final decision-making authority solely in US hands. The pact even went so far as to include a clause that confirmed US sovereign rights over the tracts of land on which the stations at La Palma and Puerto Obaldía stood. Beyond radio, the proposed treaty did not prohibit the US expropriation of Panamanian property, which had been justified under the terms of its 1903 predecessor. The proposed agreement explicitly confirmed full US authority over all Panamanian air traffic; as with radio, this stipulation hoped to formalize a concurrent American claim that even though the 1903 treaty did not explicitly address aviation, that accord's broad terms justified the claim to such authority. Finally, in a whole new insult to Panamanian independence, the treaty included a clause obligating Panama to participate in any future war alongside the United States, while giving the US military control over any operations on Panamanian soil. Whatever emergent Good Neighbor instincts informed the motivations and approaches of American diplomats, the commitment to applying those instincts was insufficiently strong to produce an accord capable of addressing long-standing Panamanian grievances. This effort, launched to resolve US-Panamanian disagreements over radio, only managed to further widen the divide and further inflame the very anti-American nationalism this accord was intended to quell.[31]

How did an initiative launched with the purported intent to account for Panama's perspectives fall so far short of the mark? The reasons are twofold. First, the State Department was ultimately complicit in the Navy's refusal to meaningfully consider Panamanian perspectives and requests. In the face of Navy recalcitrance, the diplomats largely subordinated their instincts, deferred to military demands, and advanced proposals to Panama that largely reflected Navy priorities. Second, confronted with US inflexibility in the face of domestic political pressures to get a deal done, Panamanian negotiators accepted the ultimately unacceptable terms. Even though the proposed agreement preserved and even expanded upon the broad veto powers enjoyed by the United States, the Panamanian diplomats hoped that the treaty's language of cooperation and consultation provided for a meaningful appearance of equality and shared sovereignty.[32]

While Panamanian negotiators may have felt hamstrung into accepting toxic terms for the sake of having a deal in hand, the National Assembly refused to be complicit and announced the offensive treaty would not even be accorded a vote. Then, just days after scuttling the agreement, Panama surprised US officials in February 1927 by announcing it had reached a new radio agreement with United Fruit. This new contract, once again negotiated in the absence of direct consultation with US authorities but contingent on US approval, permitted the Tropical Radio and Telegraph Company, a United Fruit subsidiary, to build another radio station. This latest station, in conjunction with a corollary railroad line, would, in part, support the company's operations on recently acquired land in Tonosí. Located near Panama's Pacific coast on the southern end of Panama's Azuero Peninsula, the site was more than a hundred miles from the Canal. The agreement also permitted Tropical to use the station to provide a public commercial radio service for sending messages to more distant points over United Fruit's larger international radio network. Only the Navy offered such a commercial service in Panama to that point. The Tropical contract ignored the 1903 treaty and the 1914 decree, and instead stipulated that its terms would be contingent on any *future* agreement signed with the United States.[33]

Striking this deal was a clever but risky move on the part of Panama's Foreign Office. American diplomats typically advocated for and supported United Fruit's efforts to negotiate radio agreements and build stations in Latin America and the Caribbean. Panama effectively dared US officials to formally protest the contract. Such a protest would force its cancellation at great cost to this "chosen instrument" and otherwise reliable corporate radio ally to the US government. At the same time, by agreeing to a contract with a company so closely identified with US policy, Panama's Foreign Ministry knowingly antagonized Panama's anti-Americanism. United Fruit was, after all, a US-based multinational company with a checkered history of widespread abuses and excesses in Panama and elsewhere in Latin America and the Caribbean. The company exercised a form of corporate colonialism by imposing exploitive and often racist labor policies that appealed to the sensibilities of its predominantly white and American-born managerial force while in the process extracting wealth from the extensive lands it controlled. Such approaches, in turn, often fueled labor unrest and anti-Americanism across the larger region.[34]

The contract was, as Minister John Glover South noted, "violently opposed here by the anti-American element." To counter that opposition and cast the contract as a symbol of Panamanian sovereignty, the Foreign Ministry shrewdly leaked South's correspondence to the Panamanian press. In those documents, South expressed his concerns to Panamanian officials about the agreement's disregard of established US authority over radio in Panama. The leak triggered a spate of hostile editorials that characterized South's objections as indicative of a "unilateral American point of view" that the United States consistently applied to all disputed interpretations of treaty rights. In objecting to South's meddling, these missives ended up defending the contract by calling for, as one such essay put it, "united resistance to the attacks of the foreign enemy." Remarkably, support for proceeding with the United Fruit subsidiary's contract became emblematic of Panamanian nationalism challenging US overreach. South, in turn, convinced Washington not to protest the treaty. Cancellation, he argued, posed the greater risk to overall American interests than the narrower challenge to US authority posed by a radio contract signed with a trusted US company.[35]

The Foreign Ministry's maneuvering further exacerbated the growing fissures between State and Navy over radio in Panama. It exposed competing interpretations of US national interest as each fought for primacy over US radio policy in Panama. In contrast to Minister South's caution, the Navy barked for an "energetic protest," as the Canal Zone's Naval Commandant Rear Admiral George Day put it. Day reiterated the now-tired arguments of inevitable interference that the station, despite its distance from the Canal, was sure to generate. The Navy, unhappy with the encroachment on the commercial services it already provided, hyperbolically insisted that United Fruit had aspired to "complete control and dominance" of radio in Panama at the expense of the Navy. United Fruit was, the Navy insisted, undermining US national interests in Panama. United Fruit's subsidiary Tropical Radio, in turn, weighed in with a forceful counter to these accusations. Its rebuttal framed the contract as an obvious benefit to US national interest consistent with its other Latin American initiatives. In urging the State

Department to allow the contract to proceed, the company touted United Fruit's record of "trustworthy cooperation" across Latin America and the Caribbean that showed the company to be "conspicuously loyal to the interests of the American government," while also expressing frustration with the obstacles it confronted in Panama that were wholly a product of US policies.[36]

For Panama, waging this battle was important for appearances and maintaining legitimacy within the Panamanian political climate. In reality, however, the prospect of Panama securing control of radio outside the Canal Zone had no prospect of meaningfully altering the American preponderance of power. If successful, Panama would, to be sure, secure a modicum of control over the development of broadcasting in the country. However, the internationally expansive telecommunications networks into which Panama sought integration through control of its own radio policies were a different matter. Telecommunications along with integrated global transportation networks surrounding Panama had been built and structured to serve larger US economic and strategic interests. Such networks, in turn, entangled large parts of undeveloped Latin America and the Caribbean with the demands and pressures of the US industrial and consumer economy. Moreover, Panama had additional disadvantages. Beyond the direct treaty-based authority American officials exercised over Panama, the United States also made the US dollar interchangeable with Panamanian currency; that move further entangled the Panamanian economy inextricably with the ebbs and flows of the American economy. Regardless of the radio dispute's outcome, Panama's position as a source of raw materials and foodstuffs heading to the United States and other more highly developed markets was never in question. The prospect of achieving more balanced economic and political development in Panama outside of the American shadow remained remote. The domestic political importance of appearing to challenge the United States notwithstanding, Panama's control over its own radio communications alone would not be sufficient to change this trajectory.[37]

With little real immediate risk posed to the United States' dominant position in Panama, the Navy's willful provocation of Panamanian nationalism and anti-Americanism appears even more egregious. The State Department certainly recognized as much. With Minister South expressing his reservations about the consequences of cancellation for United Fruit and "the possible serious injury they might suffer," State ultimately dismissed Navy's frenetic objections. Instead, it permitted the contract to move forward. In effect, Panama's Foreign Ministry and the US State Department both looked to United Fruit, albeit for very different reasons, as the vehicle to achieve their separate visions of their respective country's national interest as it pertained to radio. State's preference for pursing a more cooperative US-Panama radio policy as best suited to protecting US national interest appeared to gain the advantage over the Navy's more confrontational inclinations.[38]

However, a long-term solution to the radio dispute remained elusive in the face of a diplomatic corps not yet wholly unified behind that cooperative vision. The American Chargé d'Affaires Benjamin Muse's tense negotiations with Panama's frustrated foreign secretary (and future president) Juan Demóstenes Arosemena Barreati in 1929 punctuated that point in a very unneighborly manner. In one such meeting, an angry Arosemena lambasted US radio policy as "extremely embarrassing" for his country, depriving Panama of "the benefits of modern science and civilization," and amounting to nothing more than another "unjustifiable imposition of a great power upon a weak and helpless state." Muse dismissed the criticisms as the ravings of an unhinged diplomat, "a characteristic emotional outburst, of a kind I have become accustomed." In Muse's telling, Arosemena returned to "an unusually reasonable mood" once he vented his frustrations. This condescending dismissiveness of his counterpart's viewpoint did not offer much hope for reaching a mutually acceptable resolution.[39]

With negotiations stalemated into the next year, Panama's patience finally ran out. On December 29, 1930, President Florencio Harmodio Arosemena Guillén (no relation to the foreign secretary) informed the United States that he was revoking the 1914 decree effective immediately. Panama had signed the 1927 International Radio Convention, as did the United States. That multinational agreement bound the signatories to ensure their respective country's radio systems adhered to specific standards and stipulations that promised to minimize interference between different national systems and thereby facilitate the smooth functioning of the medium worldwide. This treaty, the president insisted, had superior international legal standing over a presidential decree singularly issued without National Assembly approval more than fifteen years earlier. Arosemena, however, did not announce the revocation to the public. In keeping it secret to all but US officials, he left room for continued negotiation. This strategy avoided the political risks that could come with announcing the cancellation of the 1914 decree to much fanfare, only to subsequently reach an agreement that might leave the United States with some unpopular vestiges of radio authority. In fact, just one month after the decree's quiet abrogation, President Arosemena was ousted in a popularly supported coup amidst ever escalating disaffection over poor governance and high-handed US policies toward Panama. His successors, however, also kept knowledge of the abrogation confined to the executive branch in the hopes of finding that still-elusive common ground over radio without fueling unrealistic public expectations.[40]

It was not to be. The Navy, for its part, was unrelenting. One such proposal from the Navy even went beyond the excessive stipulations of the abortive 1926

Alfaro-Kellogg Treaty proposal; it demanded a monopoly over all mobile radio traffic in Panama as part of any new radio agreement. Despite its reservations, State advanced it for consideration, only to be greeted by the all-too-predictable Panamanian rejection. Though State ultimately forwarded the Navy's demands, Latin American specialist Allen Dawson skewered the Navy for making proposals "distinctly onerous to Panama which do not seem essential to insure the United States an adequate measure of control." Assistant legal advisor William R. Vallance implored the Navy to simply "trust the State Department to deal with the delicate matter of negotiations and get the best results obtainable." No good, Vallance continued, could come from deliberately antagonizing Panamanian nationalism, which was committed to resisting further US encroachments. Even if Panama agreed to US demands to maintain or possibly increase its control over radio, the consequences could very well be "the possible overthrow of the Panamanian authorities who granted it." These were not idle warnings, given President Arosemena's recent fate at the hands of a coup. But the Navy stood its ground, President Herbert Hoover was disengaged, and the State Department, its reservations notwithstanding, stood aside. Consequently, Panamanian grievances at the American radio policies continued to fester and multiply.[41]

Amidst the stalemated negotiations, Panama signed yet another contract without consultation. In so doing, Panama further widened the growing State-Navy fissure. This time, Panamanian officials looked to All America Cables, a subsidiary of the International Telephone and Telegraph Company. This agreement permitted the construction and operation radiotelephone stations within Panama. The proposed stations would be incorporated into ITT's larger Latin American communications network, which provided direct links to North America and Europe, and on which many American commercial interests elsewhere in Latin America relied. The State Department supported the contract, while the Navy demanded it be blocked.[42]

The Navy, however, discredited itself in its desperate effort to torpedo this latest contract. It did so by reversing its staunch opposition to United Fruit's Tropical Radio company from four years earlier, now claiming Tropical was a "100% American enterprise" in contrast to ITT's alleged foreign loyalties. Levying the charge of anti-American loyalties against ITT due to the presence of foreign-born members on its board of directors was a favored but ineffectual tactic used by ITT's opponents; the allegation had been decisively rejected years earlier by the Federal Radio Commission. Secretary of State Henry Stimson challenged the Navy to produce substantive evidence backing its discredited claims. Instead, Secretary of the Navy Charles Frances Adams admitted that his motivations to block ITT reflected a years' old personal grudge he held against the company's chairman Sosthenes Behn. The Navy's position, Assistant Secretary of State Francis White concluded, was "perfectly absurd," "high-handed, narrow-minded," "purely arbitrary," and based on "no valid military reasons." The Navy's argument, White concluded, seemed to rest on the premise that "anybody who does not agree with them is a not a loyal American citizen and is working against the US government." Stimson had heard enough. In November 1932, he approved the All America contract over strenuous Navy objections.[43]

Stimson left office anticipating a new era of US-Panamanian cooperation girded by radio. At the end of February 1933, just prior to the change of presidential administrations following Franklin Roosevelt's election victory in 1932, the outgoing secretary of state and his Panamanian counterpart Foreign Minister Arosemena celebrated the long-awaited opening of the Tropical Radio station authorized by the 1927 contract. They exchanged messages, both of which struck the familiar notes linking radio to the furthering of international peace and understanding. The new US-Panama radio connection, Stimson effused, "was bound to decrease the possibility of misunderstanding" by facilitating warm and friendly conversations. Arosemena concurred, expressing confidence that the radio link promised to "remove any possibility for misunderstanding that might exist between our peoples." It would complement the Panama Canal, itself "so important for the benefit of mankind," while ensuring the United States and Panama would "be brought so much closer together by this service now being inaugurated." Stimson, echoing those sentiments, professed his certainty that "this added touch of intimacy can only intensify the cordial relations already existing." It would be left to the incoming Roosevelt administration to realize those aspirations.[44]

"An Outrage to Our People": The Hull-Alfaro Treaties of 1936, US Domestic Politics, and the Broken Promise of the Good Neighbor Policy in Panama

At first glance, progress toward resolving the long-standing US-Panama radio dispute and other grievances under Franklin Roosevelt seemed relatively swift. After decades of stalemate, it took just three years for American and Panamanian negotiators to reach agreement on four accords aimed at addressing a wide range of Panamanian grievances, including radio. To be fair, the sense of urgency to reset US-Panamanian relations was greater than it had ever been by the time Roosevelt took office. The Great Depression that began in 1929 and helped propel Roosevelt into the presidency also added further incentive to mitigate US-Panama tensions. The dramatic reduction in Canal traffic and trade that accompanied the global economic catastrophe disproportionally slammed Panama's underdeveloped economy. These struggles further strained relations with the United States as Panama's economic tailspin yet again underscored its subordination and vulnerability to the United States. An earnest effort to resolve outstanding issues and reduce appearances of subordination and vulnerability offered a potential corrective.[45]

Roosevelt helped set the tone by personally meeting with President Harmodio Arias Madrid in October 1933. The ensuing discussions prioritized the radio conflict. While in Washington, Arias reiterated his and his country's longstanding objection to "the indignity of not being able to control commercial radio in Panama when the defense of the Canal was not involved." Although the Navy remained strongly opposed to any concessions that undermined its smothering authority, Roosevelt's sympathies, like those of his State Department, clearly lay with Panama. The outcome of the summit was a joint memorandum released under both presidents' names that called for radio oversight on the entire isthmus to be administered by a supervisory board with equal Panamanian and US representation. The Americans further pledged that "no special restrictions [would] be placed on 'ship to shore' service with the exception of that relative to transiting the canal." This cooperative spirit endured. Confronted with the Navy's continued recalcitrance against any concessions at year's end, Roosevelt underscored his conviction that "we must carry through the intent of what we told the President of Panama when he was here."[46]

Roosevelt's inclinations emboldened the State Department. Following the Arias-Roosevelt summit, Secretary of State Cordell Hull instructed his subordinates to revisit the entire history of the US-Panama radio dispute. In April 1934, Hull assured Panama's secretary of foreign affairs Arosemena that State's reassessment examined "sympathetically Panama's viewpoint on all these matters." In fact, that reexamination produced a scathing ten-page rebuke of the Navy that had already been forwarded to the president a month earlier. The memo lambasted the Navy's repeated inclinations to prioritize its own convenience by insisting on unnecessary radio restrictions that lacked military justification. It railed against the Navy's propensity to make unreasonable demands that "antagonize Panama by continuing to treat her in a high-handed and arbitrary manner." The Navy was derelict in not recognizing that Panama's "good will and cooperation" was an essential ingredient to any viable solution.[47]

The leveling of such a blunt and unforgiving attack against a fellow department in the executive branch was unusual. The memo reached the desk of Assistant Secretary of State Sumner Welles, who was tasked with deciding if it should be advanced to the president. Given the harsh criticisms directed at the Navy, the State Department's Office of Coordination and Review, responsible for (as the name indicates) managing interdepartmental relations, cautioned

Welles to read the missive carefully before approving it. But career officials from both Latin American Affairs and the Treaty Division all lined up in support of the censure. Moreover, Welles himself was a longtime Latin American affairs specialist who had led the Division of Latin American Affairs in the early 1920s. In that capacity, he advocated pursuing diplomatic approaches toward Latin America that were exemplary of that era's nascent Good Neighbor inclinations. He was also the president's personal friend. Against the backdrop of strong departmental support, his long-standing inclinations to pursue more cooperative approaches to diplomacy with Latin America, and his relationship with the president, Welles had no reservations about forwarding the document to Roosevelt. In so doing, Welles expressed his full concurrence with the substance of the "vigorously worded" report that he believed accurately conveyed the Navy's "indefensible" attitude and approaches.[48]

This department-wide hostile consensus against the Navy found further expression when one month later Panamanian officials submitted to the State Department their own critical assessment of the Navy. The Navy, this missive insisted, was primarily culpable in perpetuating the US-Panama radio dispute. The Panamanian broadside was, in the words of one State official, a "very good" synthesis of the years-long radio dispute and required no correction or refutation. Having found its backbone, an increasingly assertive State Department refused to entertain, much less advance, any of the Navy's most outrageous proposals. Frustrated Navy officials demanded that Roosevelt intervene. Navy hopes that Roosevelt, a former assistant secretary of the Navy under Woodrow Wilson, might side with it were quickly dashed. At a July 1934 interdepartmental meeting, the president counseled that Navy demands for control over radio went beyond basic needs to assure the defense of the Canal, and "when such protection was assured, we also should give full consideration to Panama's rights as a sovereign country."[49]

Rising pressures to strike a radio deal sooner rather than later also emanated from Panama. In October 1932, the news finally leaked to the wider Panamanian public that the 1914 presidential decree had been rescinded two years earlier and since then Panama had not recognized US authority over radio in the country. With the legislature poised to pass legislation revoking the decree, the Arias administration acknowledged that the decree had already been unilaterally abrogated two years earlier and such legislation was therefore unnecessary. The now-public knowledge that the decree was already voided limited Panama's maneuverability. It could not accept any new agreement that might restore a semblance of authority to the United States within Panamanian jurisdictions. Grievances over the American role in stifling broadcasting's development in Panama had also been escalating. By the mid-1930s very little had changed since nearly a decade earlier, when the *Star and Herald* compared the state of radio development in Panama to what one might expect "in the heart of darkest Africa." It was only in August 1934 that Panamanian officials had their first opportunity to address a Panamanian audience directly over the radio. However, their platform was not a Panamanian station but rather a US-based shortwave broadcast by NBC. Soon after that broadcast, unauthorized stations began to dot the Panamanian landscape and interfere with Navy stations. In some cases, the interference was deliberate and intended to pressure the Americans further. When Station HP5A—the so-called Voice of Panama—started broadcasting in October 1934, Minister Antonio Gonzalez admitted to Washington that the interference was intentional and the Americans should not expect any corrective action from Panama until a radio agreement was struck. The State Department understood Panama was not bluffing and proceeded with negotiations accordingly.[50]

The diplomats moved quickly and had four treaties in hand by March 1936. The State Department had every reason to believe it had achieved its goal of recalibrating US-Panama radio relations on a mutually acceptable basis. The General Treaty revised some of the most offensive and controversial stipulations of the original 1903 treaty. This agreement ended Panama's status as a US protectorate. It denied the United States the right of eminent domain to seize land in the Canal terminus cities of Colón and Panama City. The accord permitted Panamanians the right to freely move through the Canal Zone. In a clear corrective to the controversies that swirled around American claims over the radio stations at Puerto Obaldía and La Palma, the main accord also stipulated the exercise of joint sovereignty with the United States over any future land acquisitions deemed essential for the Canal but that lay outside the Canal Zone. Two conventions specific to radio were subordinate to the overall General Treaty. One transferred those specific stations and the surrounding land to Panamanian control. A more encompassing Radio Communications Convention recognized Panama's jurisdiction and authority over radio in the republic, and established parameters for cooperation between American and Panamanian jurisdictions to minimize potential interference. A final subordinate agreement pledged the United States and Panama to cooperate on building the Trans-Isthmian Highway; in so doing, it abrogated the Panama Railroad Company's controversial monopoly over the right to build roads in Panama, which Colombia granted well prior to Panamanian independence. The radio and highway conventions thus addressed the Panamanian authorities' longstanding grievances over US-imposed monopolies over radio and roadbuilding blamed for hindering the government's ability to integrate the interior populations into a truly national community.[51]

The agreements were initially well received in both Panama and the United States. Their fates in the two countries, however, were starkly different. In Panama, support came from across the Panamanian upper, middle, and labor classes. After their 1936 unveiling, editorial praise celebrating the accords as substantive realizations of Good Neighbor aspirations flowed freely from the prominent Panamanian newspapers *El Panama America* and *La Nacion*. While not unanimous by any means, positive reactions overshadowed more critical assessments offered by papers such as *La Estrella de Panama*, which described the treaties as offering little more than "cultured diplomatic phraseology." In the wake of the predominantly positive response, Panama's National Assembly overwhelmingly ratified all four accords by a resounding twenty-seven-to-four vote before the year's end. Comparably widespread positive reactions also greeted the treaties across the United States in big and small newspapers alike, also overshadowing the smattering of more critical assessments. The treaties served as a form of "national repentance," the *New York Herald Tribune* noted in a typically effusive appraisal. It described the desire to improve US-Panama relations "as safe as it is wise." That consensus support notwithstanding, more than three years after their submission, the treaties had yet to receive a vote from the US Senate. Ultimately, only the General Treaty and corollary highway accord advanced to the floor of the Senate in July 1939. The United States never ratified the radio treaties.[52]

What went wrong? The Navy, with help from allies in the Army, had harnessed US domestic politics to regain what it lost at the negotiating table: the upper hand over the State Department. The US constitutional stipulation that mandated a two-thirds supermajority vote in the US Senate to ratify treaties gave the Navy its leverage. The main obstacle to ratification, the otherwise supportive *New York Times* cautioned, was "the inevitable cavilings of legislators, some of whom are apt to look upon any treaty bettering the relations between the two countries as a 'base surrender,' a 'betrayal.'" This prescient warning found expression in the willfully ignorant and remarkably dishonest critiques of the treaty like the one put forward by the *Green Bay Press-Gazette*. Treaty revision was not necessary because Panama "never even intimated that America has abused or misused its control over Panama's sovereignty," the *Gazette* contended with incomprehensible obliviousness. To play on such sentiments and keep at least thirty-three of ninety-six senators in the opposition, the Navy pursued a two-pronged offensive. First, with some assistance from like-minded Army allies, it launched a public attack on the treaties as looming threats to US defense preparedness. Second, it capitalized on domestic opposition to the treaties by recruiting influential senators most likely to share their concerns and thereby keep the treaties from securing that sixty-four-vote threshold of support.[53]

The attack against the treaties began in spring 1936, soon after their submission to the Senate Foreign Relations Committee for initial consideration. In April, both Army and Navy leaders publicly challenged the State Department by declaring their determination to, if the treaty was ratified, simply ignore any disagreeable restrictions on their freedom of action in the name of prioritizing Canal safety and defense. Headlines blared across the country that the military had "give[n] notice to the State Dept" of its intent to "hold the safety of the Canal ahead of treaty." The military would continue "to guard the Canal despite [the] pact," and will always do so "no matter what diplomats do" since the treaty was a "peril in wartime." The military doubled down on its essentially unconstitutional messaging when two prominent retired leaders, each billed as a "noted authority" on defense matters, lit into the treaties with lengthy and scathing nationally syndicated editorials published in May. Rear Admiral Yates Stirling Jr., who retired only days earlier, ridiculed the "altruistic" sentiments of Good Neighbor policies as a threat to US strategic interests. Retired Brigadier General Henry J. Reilly painted a hypothetical picture of a canal under siege by foreign enemies in a possible future war, Panamanian rabble rousers determined to undermine the United States, and a neutered US military unable to respond to either threat. The Army and the Navy both needed to preserve, as Reilly put it, the "right to take whatever steps would be necessary" as originally guaranteed by the 1903 treaty. The strategy was well suited to amplify concerns such as those articulated by conservative columnist James T. Williams Jr., whose early April column raged that "the Department of State has apparently taken the ground that the defenses of the canal are little if any part of the business of the War and Navy Departments."[54]

The Navy captured the support of two influential Senate Republicans with bipartisan appeal. California's Hiram Johnson won reelection to a fourth term in 1934 with more than 94 percent of the vote after running with the Republican, Democratic, and Progressive Party endorsements. He was a longtime proponent of a unilateralist approach to foreign policy and often used his perch on the Senate Foreign Relations Committee to oppose treaties that he believed unwisely constrained US freedom of action. The Panama accords, in his view, were just such agreements and "an outrage against our people." Maine's Wallace White, who also sat on the Senate Foreign Relations Committee, was a noted expert on communications law who had his hand in every aspect of radio regulation since 1919. More content than the showboating Johnson to work behind the scenes, his advice was sought by Republicans and Democrats alike on radio matters that were before the Senate. The respect White commanded did not bode well

Figure 13. Wallace White. Courtesy of Harris and Ewing Photographs, Library of Congress Prints and Photographs Division.

for the radio treaties. The Maine senator especially distrusted Panamanian competence to exercise authority over mobile wireless services. Intimating that military opposition had been improperly silenced, White suggested in 1939 that if military communications experts were allowed to "freely express themselves" on granting Panama such authority, they would most certainly "condemn it without reservation." Roosevelt had, in fact, dictated to his military leaders what they could and could not say when called to testify before the Senate.[55]

In July 1939, the State Department finally broke the treaty logjam, albeit only partly. It did so by leveraging an escalating fiscal crisis in Panama to secure the long-awaited ratification of the General Treaty. That accord revised the formula by which US rent payments to Panama were calculated to account for the devalued Depression-era US dollar. The State Department, however, refused to authorize any further payments until the requisite treaty with its new formula was ratified. Consequently, by mid-1939 US payments to Panama, part of the Panamanian government's indispensable revenue, were several years in arrears. The stoppage of payments further destabilized an economy already in shambles. With war clouds gathering in Europe and July marking the second anniversary of Sino-Japanese War's eruption in Asia, the lack of ratification also undermined the coordination of US-Panama defense planning. By July 1939, enough senators saw a further destabilized Panama in the context of an increasingly unstable world as the greater threat to the Canal than the General Treaty. In this context that main treaty secured sixty-five affirmative votes to ratify the main accord. Hiram Johnson, however, remained unmovable. He argued just a day before the ratification of the General Treaty that "Panama has never been of any help to the United States and I venture to say it never will."

That sentiment doubting Panama's cooperative capacities, however, spelled doom for the radio accords. Wallace White, whose credibility on radio policy was unrivaled, remained adamantly opposed and in so doing kept the radio treaties from reaching the necessary sixty-four-vote threshold. Both radio treaties never secured the votes necessary for ratification, either in that session or future ones.[56]

It did not matter. The Navy's desperate attempt to freeze in place US radio authority was too little and too late. By the time the General Treaty reached the floor of the Senate in July 1939, Panama touted twelve broadcasting stations that could reach an estimated 26,000 receiving sets. For a country whose population was 560,000, such numbers might not seem especially impressive. They do, however, point to a rapidly growing market in which ownership increased by more than 300 percent in just three years. In the absence of a jointly ratified radio agreement, these trends seemingly risked the type of endemic interference that the Navy long claimed it was working to prevent. But the Roosevelt administration was prepared. Anticipating the likely failure of the radio accords, by July 1939 the State Department had already drafted an executive order for the president's signature that enacted the main radio convention's terms within the Canal Zone. That move brought the Canal Zone into sync with Panama's oversight of radio in the surrounding country. It was an ironic exercise of presidential authority in Panama that the Navy first advocated as the means to impose US radio policy on Panama two decades earlier, only now to see it applied to contravene its own radio priorities. The transfer of ownership to Panama of the La Palma and Puerto Obaldía stations was even easier to effect since it only required existing bureaucratic procedures to transfer property from the United States to Panama. "I anticipate very little real change in the fine relationship we now have with the Panamanians," a perhaps overly optimistic embassy official in Panama informed Washington.[57]

That latter assessment was indeed far too sanguine. By 1939 Panama was, as historian John Major put it, in the throes of "the most assertive nationalism the country had yet experienced," much of it directed in the form of hostility toward the United States. The exceedingly long delay in the US ratification of the General Treaty, the outright scuttling of the radio treaties, and the public insults lobbed at Panama and its integrity to justify treaty opposition predictably helped intensify that anti-American nationalism. In a blow to their domestic political standing, Panamanian leaders were put in the humiliating position of having to respond over the course of three years to the harsh public rebukes from outspoken treaty opponents by providing their own assurances to a US audience that Panama was indeed faithful to its commitments. The treaties, or at least the American fight surrounding them, managed to produce the very outcome they were supposed to mitigate: a further poisoning of US-Panamanian relations.[58]

The resolution to the radio dispute remained forever elusive. An "extremely informal arrangement," as one State Department official put it, coordinated radio operations across the American and Panamanian jurisdictions in the country into the late 1940s. By September 1947, it was an arrangement that depended entirely on the continued services and good health of the elderly Richard D. Prescott. Prescott, in fact, was the founding chairman of the Panama Radio Club. It was his signature affixed to the letter sent two decades earlier that had unsuccessfully pleaded with Canal Zone authorities to permit amateur broadcasting. By 1947, Prescott, then a reserve officer in the US Army who had reached the rank of colonel while on active duty through the end of World War II, was Panama's long-serving telecommunications advisor. He was tasked with coordinating the radio operations of the two separate jurisdictions in the absence of formal agreement. Though the arrangement "appears to have worked satisfactorily," State Department communications expert William O'Connor reported in September 1947, "it should also be remembered that Mr. Prescott is an old man." Retirement or death loomed as realistic near-term possibilities, O'Connor warned. In the event of either outcome, there was no guarantee Panama would replace Prescott with another American appointee or one of comparable competence.[59]

With yet another draft radio treaty under consideration in autumn 1947, O'Connor warned against requesting changes. Any requests to dilute terms ceding authority to Panama promised to provoke Panama's fierce opposition, as had been the case so many times before. O'Connor's cautionary advice notwithstanding, the familiar grumbling and insults from outspoken American critics followed. An angry Panama then withdrew from the negotiations and proclaimed it would no longer entertain any bilateral radio agreements with the United States. Instead, in October Panama signed a multilateral telecommunications agreement with seventy-seven other signatories. Multilateral, not bilateral, agreements, Panamanian authorities decided, had to remain the basis of US-Panama radio cooperation going forward. Only then could Panama ensure it would be treated as one among many equals, not as a colonized afterthought. How much longer Richard D. Prescott continued in his capacity as the coordinator of radio communications across both jurisdictions is unclear.[60]

* * *

For all its success in blocking the State Department's diplomatic priorities, the Navy's narrowly conceived US-centric military vision for radio in Panama produced few long-term advantages for itself or anyone else. The Navy lacked imagination, understanding, and empathy to account for radio's transformation and evolving significance within the broader political, economic, and cultural contexts in which it functioned. It sought to retain control over radio in Panama solely for the sake of its own convenience and myopic priorities. Outside of Panama, these evident limitations of the Navy's leadership had in the aftermath of World War I squandered much of the Navy's once-preponderant influence over US radio development and policy.[61] While the unique position that the Panama Canal occupied in American economic and defense calculations helped sustain the Navy's influence over radio in Panama as it diminished elsewhere, its stunning ineptitude in radio matters ultimately undermined US-Panamanian relations without preserving the modicum of control over radio that the Navy long insisted was vital to US security.

This tone-deaf and often incompetent execution of radio policy cast a long shadow across the remainder of the twentieth century. Academic and popular histories of Panamanian radio still single out the United States for having unfairly stifled radio's development. Hector Staff, a Panamanian radio scholar, argues that broadcasting on the isthmus owed its 1930s "clandestine" beginnings to the "rigorous control" and "constant pressure" imposed by the United States, which Panama ultimately overcame in pursuit of the "sovereign motivations of the people, in favor of our nationality." In the late 1980s, the United States proposed to launch a shortwave broadcast service intended to transmit programs into Panama for the purpose of promoting democracy. In response, *La Estrella de Panama*'s editorial board evoked the original US-Panama radio dispute by claiming that Americans sought "once again to repeat their colonialist tendencies" over "our national radio communications system." Even present-day Panamanian websites offering brief nonacademic overviews of Panama's "humiliating" radio history frame the medium's development as a triumph over US obstacles and American claims that Panama's radio frequencies "belonged to the United States." The notion of airwaves as tangible territory endures, as do the convictions that the United States had unjustifiably colonized those airwaves in tandem with Panamanian territory more broadly since the earliest days of Panama's independence.[62]

Radio concessions, in fact, posed no immediate risk to meaningfully altering the imbalance of power that defined the US-Panamanian relationship. Resistance to such concessions instead reflected a larger enduring pattern of misguided assumptions, missteps, and actions that over time increasingly undermined the ability of the United States to maintain and exercise effective authority in Panama. Such shortcomings contributed to the repeated outbreak of violent and often deadly riots from the 1940s through the 1970s. These riots included one that erupted soon after Panama withdrew from the 1947 negotiations yet again seeking a radio treaty; the spark was concurrent American military efforts, again over State Department reservations, to strong-arm Panama into agreeing to long-term leases to military bases while refusing to entertain Panamanian requests for joint control. This dysfunctionality in the US-Panama relationship ultimately culminated in the 1977 treaty in which the United States finally agreed to relinquish its authority and fully cede control of the Canal to Panama before the end of the century. Amidst the instability and increasing toxicity in the overall relationship, US officials saw no alternative.[63]

American opposition to Panamanian radio aspirations in the period under consideration was, to reiterate the 1932 assessment by the State Department's Francis White, perfectly absurd. Ultimately, Panama's assertion of control over radio in its jurisdictions over US objections never provoked the Canal-incapacitating communications Armageddon that the Navy had long threatened. Practical calculations reigned supreme. It was one thing to use interference as a tool to wrest control from the United States. Once that control was in hand, Panama's development of its domestic radio networks and markets, including broadcasting, dictated minimizing interference. In exercising authority over radio, it was therefore in Panama's interests to adhere to larger international radio standards and practices, regardless of whether the binding agreement was a bilateral US-Panama pact or a broader multilateral accord that pledged both the United States and Panama to the same terms.

Might an American willingness to account for Panamanian nationalism and make concessions to it on matters like radio have made a difference in the longer term? Could doing so have meaningfully altered the debilitating century-long intensification of anti-American Panamanian nationalism? If there is any merit to such far-fetched speculation, it certainly would have required substantive adjustments to policies and approaches far beyond radio. But if radio is any indicator of that larger picture, the Navy's woeful inability to "imagine what went on at the other end of the cable line," as Latin America specialist and diplomat-turned-historian Dana Munro once put it, was the problem. That shortcoming was the crux of the American failure to conceive and execute a radio policy toward Panama that effectively served either State's or Navy's conception of the US national interests. Instead, the radio policies pursued over a span of three decades from the 1910s into the 1940s primarily served to magnify and help intensify the far bigger animosities that long plagued the unequal US-Panamanian relationship.

CHAPTER 6

"An Almost Unbelievable Disregard of the Interests of the United States Listeners and Broadcasters"

US-Cuban Relations, American Identities, and the 1946 North American Regional Broadcasting Agreement

Farmers and growers in the American Southwest were stunned. An international radio accord signed at the beginning of 1946 was forcing the high-powered Los Angeles-based radio station KFI to share what had been its proprietary frequency assignment on 640 kilocycles (kc) with a distant but powerful Cuban station. The news infuriated farmers and growers throughout the agricultural Southwest. They relied on KFI's weather reports during the winter and early spring to protect valuable but vulnerable crops from looming inclement weather, especially frost. Protests flooded the US State Department. Typical was the complaint from H.S. Stewart, vice chairman of the San Bernardino County Farm Bureau's Citrus Department. He insisted that Cuban broadcasts on 640kc would undoubtedly interfere with KFI. Such interference would undermine the ability of any farmer or grower living more than seventy-five miles outside of Los Angeles to hear those all-important weather reports. Acquiescence to Cuban radio demands by US negotiators was therefore "a deliberate sacrifice of California and Arizona agriculture to a doubtful policy of good will toward neighboring nations," Stewart fumed.[1]

In voicing this protest, Stewart joined scores of individuals and more than twenty agricultural organizations that represented thousands of farmers and growers from California and Arizona. At issue was the impending expiration of the original North American Regional Broadcasting Agreement (NARBA). That treaty, in effect since March 1941, provided the basis for radio station frequency assignments and power allocations in the northern portion of the Western Hemisphere. Its objective was to minimize cross-border interference among the signatories. Now, five years after that treaty helped shut down John Brinkley's and other Mexico-based border blasters, a NARBA revision that gave Cuba rights to KFI's clear channel frequency of 640kc provoked farmers, growers, and local leaders from the agricultural Southwest to protest the agreement. To the protesting Americans, Cuba had staked a claim to something akin to sovereign American territory.

It was a ludicrous position in the view of the Cubans. They could point to the US refusal to relinquish its claim to *actual* Cuban territory, while Cuba's airwaves faced a deluge of American broadcasts. The US-Cuban relationship had navigated complicated terrain since the US victory over Spain in the War of 1898 ended Spanish control over Cuba. For many Cubans, anti-Americanism born of the dominance the United States subsequently exerted over the island's economics, politics, and territory coexisted uneasily with admiration for American success in technology-based modernization. Cubans blamed the suffocating American presence for hindering Cuba's efforts to emulate American technological achievements for their own purposes. In that regard, the 1941 NARBA, which yielded a handful of frequencies to Mexico, confirmed the dominant American position in hemispheric broadcasting and continued to stifle Cuban radio development. In 1946, Cuba laid claim to 640kc, insisted that the expiring NARBA be revised to account for that claim, and made clear it would use the frequency whether or not NARBA was changed.[2]

At its heart, this story is one about the enduring strength of national and regional identities in a globalizing world. Cuban encounters with American radio programing inspired Cubans to fight for distinctly Cuban radio interests. Cuban success on that front inspired the infuriated farmers and growers of the American Southwest to rail against their government's acquiescence in ways that underscored an enduring regional identity within the larger American polity. Those reactions, in fact, were deeply entangled with the complex history of American identity formation, which had from the nation's earliest years privileged specific regional loyalties that coexisted alongside both local and national ones. Once again, border-crossing broadcasts antagonized existing tensions and, in turn, provoked decidedly provincial reactions.[3]

It is that privileging of regional identity among southwestern farmers that speaks to a particularly enduring theme in US history. Media scholar Susan Douglas discerns a counterintuitive pattern of Americans engaging with nationally and globally extensive communications technologies in a way that cultivated more isolationist, ethnocentric, and narcissistic impulses among its audience. This "turn within," she points out, was even evident in how radio broadcasting in the World War II era "promoted more regional identifications sometimes in opposition to such homogenized national affiliations." Such dynamics are apparent in the way agricultural producers from the American Southwest mobilized against the 1946 NARBA. Situated at the intersection of local, regional, national, and global processes, this chapter answers historian Joseph Fry's call to pay more attention to the role of domestic regionalism in US foreign relations. "Addressing the intersection of domestic regionalism and US foreign policy," Fry explains, "significantly enhances our understanding of both regional and national American history." The vehement and regionally based opposition to the 1946 NARBA demonstrates the verity of that observation. The 1946 NARBA and the concessions it offered Cuba inspired the protesters to engage long-standing *domestic* battles that had pitted different groups of Americans against one another for generations.[4]

KFI, NARBA, and Cuba

Founded in 1922, Los Angeles-based KFI was one of the powerful pioneering stations in American broadcasting history. Indicative of how the clear channel system favored commercial broadcasters, the station became an NBC affiliate upon the network's 1926 founding. KFI's 50,000-watt "clear channel" status allowed it to serve a rural audience far outside of Los Angeles. In an effort to win favorability and credibility with its growing audience, KFI's programming included civic, educational, and public service initiatives. Those efforts involved outreach to rural listeners through farm and weather reports that targeted audiences well beyond the Los Angeles area and into distant Arizona. Such reports gave KFI's rural listeners valuable advance notice to protect their crops in the face of threatening weather.[5]

The first North American Regional Broadcasting Agreement, which went into effect in 1941, embraced the clear channel framework that Canada and the United States first adopted in the Western Hemisphere. Though a primary objective of that first NARBA was to resolve the long-standing US-Mexican border radio dispute that spanned the 1930s, the resulting 1941 accord also sought to minimize cross-border interference among rapidly growing broadcasting markets in other neighboring countries as well. Like Mexico, other signatories had sought space on the American-dominated radio spectrum to accommodate their growing broadcasting markets. The 1941 NARBA thereby provided a transnational framework of expectations and standards that facilitated coexistence on the airwaves. The five-year agreement, signed by the United States, Canada, Cuba, Haiti, Mexico, and the Dominican Republic, established different classes of frequencies and power levels for different types of stations (such as those designated for entertainment, aviation, navigation, and government services). The United States, demanding recognition of its more developed broadcasting industry relative to its neighbors, initially retained priority over thirty-two clear channels. Canada and Mexico each claimed six, and Cuba received one. The treaty also established the so-called 650-mile rule as an adjunct to the clear channel system: a country less than 650 miles away from a nation holding priority to a clear channel could receive permission to broadcast at very low power levels on that frequency, so long as the broadcasts ceased in the evening hours. That evening restriction reflected the fact that nighttime atmospheric conditions enabled transmissions to travel farther distances. The rule was designed to give the smaller signatory nations with few or no clear channel assignments a bit more space to maneuver on the crowded spectrum while maintaining the intended exclusivity of those clear channels. Consequently, the 1941 treaty effectively formalized KFI's privileged clear channel position on the US radio dial within the context of NARBA's larger international framework.[6]

The Cuban government soon regretted its NARBA commitment. In 1935, Cuba had fifty-one standard broadcast radio stations. By 1941, NARBA's inaugural year, that number reached eighty-three. Confined by NARBA, however, Cuba subsequently squeezed only ten more standard broadcast stations onto its airwaves by 1945 to bring its total to ninety-three. Given that Cuba had a population of less than five million people and only an estimated 250,000 receiving sets, nearly a hundred stations might seem sufficient. However, the vast majority of those stations operated at very low power—generally 250 watts, with some as low as 100 watts—and served extremely small areas. In fact, Cuba's 1946 NARBA delegation claimed that only seventy-eight unique standard broadcast stations were operating in Cuba at the beginning of that year. The larger figure of ninety-three did not consider that broadcasters, such as "Radio Habana," simulcast their programming on different frequencies, a consequence of so many stations being restricted to using low power to serve small areas. The NARBA framework, which attempted to divvy up the frequency spectrum according to the situation at the treaty's signing, rather than anticipating growth in subsequent years, prevented Cuba from building stations with more power that might have served larger areas and audiences; meanwhile Cuba's large number of existing low power stations left little room to add any more.[7]

Cuba's determined pursuit of its international radio interests at the expense of American broadcasting became entangled in the historical complexities of the US-Cuban relationship dating back to the nineteenth century. During Cuba's rebellious last years as a Spanish colony, many Cubans enthusiastically embraced Americanization and American influences as a means of distinguishing Cuba from its "backward" colonial overlord. Seeking to emulate America, these Cubans proudly viewed their country as a beacon of modernity in an otherwise undeveloped Latin American world. However, the American victory in the War of 1898, which ousted Spain from Cuba, greatly complicated those sentiments. US policymakers and allied business interests had established alternative forms of economic and political dominance over the newly independent nation, which effectively rendered Cuba an informal American colony. That dominance included a perpetual American claim to the naval base at Guantanamo Bay and a right to intervene in Cuba if the Americans identified some threat to Cuban independence (meaning a threat to American economic and political interests). The pervasive American presence in Cuba also included the proliferation of American images of and ideas about Cuba pervasive in American media and popular culture. Such images and ideas impressed upon Cubans how their northern neighbors viewed them not as "civilized" equals in the modern world, but as primitive, exotic, and sexualized. A corresponding flood of more than two million American tourists between 1920 and 1940 confirmed for Cubans that Americans saw Cuba as a vast playground for drinking, gambling, sex, and all kinds of other illicit pleasures that contravened presumed standards of modern and civilized conduct. In light of that overwhelming American presence, Cuban nationalism predictably developed with a distinct "anti-Yankee" flavor during these years.[8]

As was the case with Mexico, Cuba's proximity to the United States—within the reception range of many American stations—underscored to many Cubans how thoroughly the United States dominated hemispheric radio. In so doing it also drew attention to the broader dominance that the United States exerted over the island. Cuba, in fact, had prided itself on the technological sophistication of its radio industry. Many Cubans viewed their broadcasting system as both a symbol of Cuban modernity and a conveyor of high culture and education. In this view, Cuban radio favorably distinguished itself from other nearby countries and especially from the crassly commercialized broadcasting practices of its dominating American neighbor. Indicative of this conviction, Cuba's 1939 radio decree specifically prohibited broadcasts from introducing to listeners "uncultured, bad customs, and other perturbations" that ran counter to "promoting improvement and education . . . to the core of the Cuban family." Radio should be, the decree preached, "an eloquent manifestation of each country's degree of progress and civilization." But NARBA instead had confirmed and perpetuated a US stranglehold over hemispheric radio, which many Cubans blamed for unacceptably stifling Cuban radio development. These lamentations corresponded nicely to an anti-American worldview that blamed the enduring reverberations of Cuba's turn-of-the-century colonization for throttling Cuban development on a larger scale.[9]

Cuban anti-Americanism endured through the 1930s and 1940s. Franklin Roosevelt's "Good Neighbor" initiatives of the 1930s, which included renouncing the right to intervene in Cuba, failed to effectively mitigate it. Just as concerns over German ambitions in the Western Hemisphere prompted US officials to recalibrate strained relations with Mexico and Panama, the United States tried a comparable pivot with Cuba. But even US acquiescence in 1940 to a new and deliberately provocative Cuban constitution seeking to roll back American influence failed to stem the anti-Americanist tide. The decision not to block the ratification of the document—despite US Ambassador George Messersmith lambasting it as "radical," "dangerous," and "utterly hopeless"—seemed to work at first. When the United States went to war against Germany and its Axis ally Japan in 1941, President Fulgencio Batista—elected under that new constitution—had Cuba declare war on the Axis coalition in exchange for lucrative American economic and military assistance through the Export-Import Bank and the Lend-Lease program. However, Batista paid a steep political price for positioning Cuba so squarely in the American corner during the war. Batista's bid for a second term in 1944 met defeat at the hands of his longtime political enemy Ramón Grau San Martín, who rode to victory on a constellation of political support built through the deliberate cultivation of Cuban anti-Americanism.[10]

Against this backdrop of simmering animosity, the 1941 NARBA's pending expiration in March 1946 offered an opportunity for the recently elected Grau government. It provided a means to remedy this radio-based example of its country's "unjust treatment" at the hands of the Americans. The Cuban delegation sent to negotiate the revisions demanded rights to broadcast over more frequencies and insisted on abolishing the 650-mile rule. The Cubans argued that, given Cuba's location in the shadow of the spectrum-hogging United States and other hemispheric actors, the 650-mile rule unduly throttled its ability to fully develop its broadcasting industry. The delegation railed against the "insuperable difficulties" this "detrimental and unnecessary" rule imposed on Cuban radio. The other signatories suggested Cuba agree to a two-year NARBA extension, but then negotiate a bilateral agreement with the United States to ease the 650-mile rule as it applied to Cuba. For Cuba, that proposal was a nonstarter. It would not extend what it deemed a hopelessly flawed agreement that gave undue preferences to the United States. After all, what were the chances that this same United States would then agree to subsequent concessions that Cuba insisted were necessary to better develop its own broadcasting service?[11]

Securing the right to broadcast over KFI's clear channel 640kc became the focal point of the Cuban government's efforts to resolve its grievances. Carlos

Maristany, Cuba's subsecretary of communications and the head of its NARBA delegation, complained that because of the "arbitrary" 650-mile rule, Cuba's request to use 640kc full time had been refused. And yet, this same frequency was subsequently assigned to low power stations in the Bahamas, as well as Newfoundland, two signatories that had only belatedly joined the 1941 NARBA. Moreover, the development of directional antennas, the Cuban delegation argued, had rendered the 650-mile rule obsolete. Such "technological progress" combined with "good engineering" would all but "guarantee the elimination of objectionable interference," Maristany insisted.[12]

With NARBA set to expire on March 29, 1946, Cubans were in a particularly strong position. At that point, Cuba would be free to broadcast on any frequency, regardless of previous assignments. Without Cuba's participation, an extension of NARBA by the other signatories was worthless. "Cuba had less to lose than any country," recalled Raymond Guy, an NBC engineer who attended the 1946 conference as a technical expert. Should the conference fail, Guy confirmed, "they could go back home, use all the power they wanted . . .

Figure 14. E.K. Jett (standing at the head of the table), meeting with airline executives in 1937 on the subject of air safety. Courtesy of Harris and Ewing Photographs, Library of Congress Prints and Photographs Division.

and just about wreck our broadcasting system." The Cuban delegation was determined to apply that leverage to maximum effect. "In short, it can be said that the Washington Conference was held almost entirely for the benefit of Cuba," reported E.K. Jett, head of the American delegation and member of the US Federal Communications Commission.[13]

The Cuban delegation arrived in Washington demanding twenty clear channel frequencies, more than the combined total assigned to Canada and Mexico. That strong opening volley was designed to appease Grau's anti-American domestic constituency. Behind the scenes, the Cuban delegation offered, as one American radio industry representative put it, subtle "grapevine hints" to its counterparts that it would settle for less. Having shown their hand in that way, the initial demand provoked unanimous and unyielding opposition from the other delegations. The Cubans quickly revised their demands downward to ten clear channels, otherwise known as Class I-A stations. That number, however, proved to be Cuba's bottom line. Five of those ten frequencies had been assigned to the United States under the 1941 NARBA. Cuba also wanted access to eight additional "Class I-B" regional channels, all of which the 1941 agreement had assigned to the Americans as well. KFI's clear channel 640kc had to be part of the package. To illustrate their determination, Cuban representatives answered other nations' objections by walking out in protest. It was up to the other delegations to accept the Cuban demands, or let the original NARBA expire.[14]

This posturing appalled the Americans. Francis Colt de Wolf, the head of the State Department's Telecommunications Division who served on the US delegation, scoffed at the "impossible" original demand for twenty clear channels. The trade magazine *Radio Daily* applauded the American delegation for its "solid wall of resistance" and "turning a deaf ear" to Cuban demands. But the Americans could only ignore Cuba for so long, given the treaty's pending expiration. Although it lambasted Cuba's "excessive" demands, *Broadcasting* magazine's editors conceded that a frequency war "of course, is undesirable, and should be avoided at all costs." And yet, the editorial board, a strong advocate of the commercialized American system of radio and the clear channel system that underpinned it, had no idea how such a war might be avoided in the face of Cuban intransigence. "We hope Cuba sees the light," was the best it could muster.[15]

Cuba, predictably, did not see the light. The revised NARBA agreement, *Broadcasting* subsequently reported, "gave Cuba every principal demand for which her delegates fought. Cuba went home victorious." Notably, the agreement was not a formal treaty. With the clock ticking toward NARBA's expiration, the signatories approved an "interim agreement," which in theory was supposed to extend the original treaty, save for mutually agreed upon changes acceptable to all parties for three years or until a new treaty could be negotiated. Framed as a temporary extension of the original NARBA, the 1946 agreement therefore did not require US Senate ratification. Yet, the new rights Cuba wrestled from its hemispheric neighbors indisputably went far beyond anything stipulated in the 1941 accord.[16]

This deviation was guaranteed to provoke controversy. Louis Caldwell, counsel to the Clear Channel Broadcasting Service (CCBS), unsuccessfully lobbied the State Department to renounce the agreement's terms. "Cuba perpetrated a radio hold-up," Caldwell fumed. Seeing his clear channel constituents' privileges being diluted, Caldwell preferred to risk a radio war on principle rather than accept Cuba's demands. The conference was a "fiasco," he claimed, and the US delegation had been "overimpressed by the Cuban threats which were a little short of insulting." Edwin Craig, owner of the Nashville's clear channel WSM and the chairman of the CCBS, concurred. He derided the "deplorable concessions" and "abject surrender." He railed that "principles vital to the sound regulation of broadcasting, internationally and domestically, were surrendered or badly compromised" just to "appease Cuba." The Clear Channel leader claimed that a clueless State Department simply "shrugged helplessly" before it "sandbagged" American listeners.[17]

Craig's sentiments, expressed publicly and repeatedly, reflected the precise type of American disregard for Cuba that had long fueled Cuban anti-Americanism. Thanks to an interconnected world, Cubans learned of such harsh criticisms soon after they were levied. For example, an editorial in *Alerta*, a popular left-leaning paper, found the "raucousness" of Craig's tirades about American surrender and appeasement especially condescending and ire-provoking. However, the paper also concluded that Craig's tantrum should inspire pride among Cubans, because it confirmed the Cuban delegation's impressive technical, legal, and diplomatic successes in Washington. Craig's offensive allegations aside, the paper concluded that "as Cubans we have to feel satisfied and proud to have as citizens such capable technicians and skilled diplomats."[18]

The US delegation did not earn similar accolades. Negotiators, however, believed that the advantages the revised NARBA offered far outweighed any concessions made. To be sure, delegation member de Wolf expressed regret over the cost of the accord, a tally that included Cuba's rights to broadcast on KFI's 640kc. The weary and sleep-deprived delegation head E.K. Jett also acknowledged it was a "victory for the Cubans," but defended the result in *Broadcasting* magazine. "It was essential that we reach a unanimous agreement," Jett explained, warning of the "eventual chaos" that would roil the airwaves if Cuba went rogue. Despite the concessions, the accord was, in fact, protecting the "wide-spread interests" of the United States. Publicly, Jett assured American listeners that the agreement was not so bad, noting "I personally feel satisfied, though not happy in all respects."[19]

Protesting for Region, Nation, and Station

KFI and its listeners, however, were neither satisfied nor happy in any respect. William Ryan, the station's general manager, condemned US acquiescence to the "exorbitant demands by Cuba," declaring that it demonstrated "an almost unbelievable disregard of the interests of the United States listeners and broadcasters." Despite KFI's clear channel designation, its broadcasts still had not been completely immune to interference from the distant 5,000-watt station in the Bahamas that had been allowed use of the frequency. But at least that outlet, as per the 650-mile rule, went off the air at 7:00 p.m. in the west, still an hour before any reliable overnight weather report could be broadcast. However, the 1946 NARBA allowed a Havana-based 25,000-watt station to broadcast continuously. The requirement that the Cuban station use a directional antenna was of no comfort. "We in Coachella Valley believe that no station, including the Bahamas station—not to mention Cuba, should be allowed KFI's frequency," protested Eugene Jervis, secretary-manager of the United Date Growers Association.[20]

Prodded by the CCBS, the Los Angeles Chamber of Commerce and more than twenty agricultural organizations, speaking for tens of thousands of farmers throughout the agricultural Southwest, joined KFI's Ryan and the farmer Jervis in flooding the State Department with protests. The California Farm Bureau and its county-level affiliate chapters, along with several farmer cooperatives, agricultural exchanges, local governments, and other interest groups from across the region, registered their staunch opposition. Claiming to protect the "special needs" of farmers under siege, the California Farm Bureau passed a resolution demanding that "clear channel stations be held inviolate by international agreement and that the power of these stations be increased." Above all, NARBA protesters feared that the economic costs of disrupting KFI's weather reports would be staggering, even devastating. "KFI's broadcasts to California and Arizona agriculture are the most important service by any station in this country due to Western climatic and topographic conditions and specialized winter production of crops," explained S. Erle Goodall, secretary of the Southern California Farm Bureau Federation. "Only by KFI's fast and clear broadcast of frost developments can our citrus, avocado, deciduous fruits [and] vegetables and even livestock be saved from serious loss."[21]

The loss of those advance warnings had both regional and national implications. The disruption of KFI, these opponents of the NARBA revision feared, threatened an already struggling regional economy with potential losses estimated at a half-billion dollars' worth of crops and livestock. In March 1946 unemployment in the region was nearly double the national rate of 6.8 percent and farm income had been declining. Nonetheless, California was the top agricultural state in the nation in terms of production and value. In 1946, California agriculture produced 8 percent of the country's food and accounted for 10 percent of the overall value of the food produced and consumed in the United States. Three-quarters of the nation's oranges and one-seventh of the nation's grapefruits came from that state. KFI served the top-producing agricultural counties in the nation: Los Angeles County was first, Orange seventh, San Bernardino eighth, Imperial eleventh, Riverside seventeenth, Ventura twenty-first, and San Diego thirty-sixth. By way of comparison to California's significance, Texas was the number two agricultural producing state overall (and had been ranked number one only a decade earlier), but placed only one county among top one hundred in 1945 (Hidalgo at thirty-one). Maricopa, Arizona, one of the more distant counties served by KFI and therefore one of the most vulnerable to possible disruptions, ranked fifteenth in the nation. Unsurprisingly, growers from each of these California and Arizona counties vehemently protested the 1946 NARBA.[22]

This interconnection of regional and national outlooks infused the objections to the 1946 NARBA. The United Date Growers Association accused Washington of being willing to "sacrifice the well being of California and Arizona farmers who produce a large share of food consumed in this nation." R.E. Badger, who presided over both the Associated Farmers of California and the Escondido Lemon Association, similarly warned that "the food supply of the entire nation" was at risk. That latter turn of phrase underscoring a national threat appeared either word for word or with slight modification in the protests forwarded by several other county-level Farm Bureau organizations and their affiliates. The flood of protests reflected the latest effort in a history of well-organized lobbying campaigns spearheaded by the Farm Bureau. The Farm Bureau was the country's most powerful agricultural lobby, with more than 1.3 million members by the end of the 1940s and a disposition toward protecting wealthier, large-scale agricultural interests, which were dominant in the Southwest. This organization had an impressive recent record of effectively organizing in response to perceived federal transgressions. During the Depression-era 1930s, California and Arizona agricultural interests, particularly the citrus growers, blamed pro-labor federal policies for their persistent worker shortages. The California Farm Bureau Federation and its affiliated lobby group, the Associated Farmers, successfully worked hand in hand to help stifle New Deal–era efforts to unionize and provide labor protections to agricultural workers. These same agricultural lobbyists, now working together with the CCBS, sought at this point to squelch the 1946 NARBA. By April, many of those who had protested NARBA in writing had traveled to Washington personally, where they provided further testimony against the agreement at ongoing FCC hearings about clear channel licensing preferences.[23]

The southwestern agricultural interests protesting the 1946 NARBA fit into a particular "Far West" regional identity in the United States. The Far West is a culturally distinct region whose eastern limits begin roughly at geographic middle of the United States. It stretches from the Dakotas in the north to portions of Oklahoma in the south, extending westward into eastern portions of California and the other West Coast states (and dipping south to include the more northerly portions of the southwestern states, including Arizona) until it yields to the culturally separate and decidedly urbanized West Coast. Amidst its challenging environment, the Far West's agriculture evolved toward large-scale, well-organized, and capital-hungry operations. As the United States industrialized in the late nineteenth and early twentieth centuries, the Far West found itself especially vulnerable to outside economic and political forces based in the urbanizing centers of the United States. Far Westerners blamed big banks, railroads, and a growing but unsympathetic federal bureaucracy staffed with impractically educated elitists for compromising access to credit, markets, and valuable agricultural land. On the eve of the Second World War, regional dispositions toward international neutrality and nonintervention further distinguished it from the purported internationalism and more interventionist dispositions associated with the urban areas of the country and the federal government.[24] That 1930s noninterventionist disposition was another manifestation of the region's rural inhabitants' perceived dependence on and vulnerability to distant and disruptive events beyond their control. For their part, the more internationalist urban-minded political and economic elites often dismissed rural Americans as ignorant and backwards, and in desperate need of progress and uplift. While this urban-rural antagonism was a nationwide phenomenon, the rural Far West's resistance to the internationalist and interventionist policies supposedly favored in Washington was higher than in other parts of rural America, while the contempt in which the Far West held the federal government was perhaps stronger than anywhere else in the country, save the Deep South.[25]

The long-standing regional disposition to distrust and loathe Washington shaped how the southwestern agricultural interests attacked the 1946 NARBA. In the postwar era, these existing animosities intensified amidst convictions that Washington willingly sacrificed domestic agricultural interests in exchange for securing international agreements deemed to be in some larger national interest.[26] A decade earlier, those who protested Brinkley, including those from the Far West, typically seized on their sense of national identity to demand corrective action from federal officials who purportedly were beholden to act in the national interest but had yet to do so when it came to Brinkley. The context surrounding the 1946 controversy was quite different. At this point, federal officials had already acted in the purported national interest by agreeing to a revised NARBA without seeking Senate approval. In so doing, they acted in a way that not only ran counter to existing perceived regional interests but played into the Far West's disdain for how federal officials exercised authority to its detriment. The 1946 NARBA accord and acquiescence to Cuba's demands for KFI's frequency seemingly confirmed their worst fears of the federal government as a remote yet highly disruptive antagonist.

As such, the accord provided an opportunity to heap more scorn on a distant, meddling, and already-despised federal government. Noting that the 1946 NARBA posed "a direct threat to the livelihood of thousands of growers," Santa Ynez Farm Bureau chairman R.E. Herdman advised Washington that (again deploying language repeated in several other objections) "we all violently protest any duplication of KFI['s] frequency." The United Date Growers Association echoed CCBS chairman Edwin Craig's accusation that the 1946 NARBA reflected nothing less than the "appeasement" of Cuba. Arthur Isham, the public relations director of Redlands, California, applied the same derogatory descriptor to the revision, adding it was a consequence of the State Department's "flagrantly short-sighted and indefensible act." This atrocious decision, Isham seethed, reflected the department's "complete ignoring of the needs and rights of US citizens."[27]

This equation of 1940s US radio policy toward Cuba with appeasement is, at the very least, ironic. By the mid-1940s, an appeasement charge carried an implicit critique of the noninterventionist and allegedly isolationist attitudes that pervaded the Far West a decade earlier on the eve of World War II. The use of appeasement by these Far West protesters to lend authority to their grievances underscores how pervasive and compelling this simplistic historical analogy had become in the postwar American political environment. Opposition to appeasement in any form had become central to postwar US foreign policy discourse, and its deployment against the 1946 NARBA was designed to strengthen the credibility and persuasiveness of the criticisms. At the same time, the opposition to NARBA underscored the strain of American foreign policy that demanded unilateralism. In this view, long-term international agreements and international organizations unacceptably bound the United States to the interests and policies of presumably lesser international actors.[28]

Thus, the problem with the 1946 NARBA's alleged appeasement of Cuba lay in the State Department's internationalist and even conspiratorial instincts to subordinate American freedom of action with regard to radio. Arthur Isham of Redlands, California, followed up his appeasement charge by describing the agreement, presented as a fait accompli, as "one more shining example of one Department doing something important and drastic without consultation . . . and without due regard for its effects on great segments of the people not directly in contact with the Department." "We have had our shirts traded off already in secret," another aggrieved farmer from nearby Mentone, California, complained. Former Redlands city council member and grower H.H. Ford charged that this sacrifice of his region's livelihood reflected the misplaced priorities of a small cabal of State Department negotiators who sought to cultivate access to "Latin American music and the good will of Havana honky-tonk owners." Another Redlands grower, James Leonard, concisely captured the Far Western, anti-Washington, and anti-internationalist sentiment that pervaded the protests when he wished that the State Department would for once just "stand with us." Reference to the collective "us" while levying such sharp criticisms spoke to a sense of a shared American identity, albeit one betrayed by actions in Washington detrimental to their regionally based concerns.[29]

To make their case against NARBA, KFI's advocates navigated through a multilayered combination of local, regional, and national identities that comprised the totality of their overall outlooks. On one level, the spark behind the NARBA protests spoke to more locally focused agricultural concerns encompassing the regional climate and radio listening habits very specific to their southwestern locale. Also specific to this southwestern subset of Far Westerners was a defensive-minded sense of identity that had long been attuned to combating and marginalizing perceived foreign threats to the agricultural economy. The California Farm Bureau, along with allies in the Los Angeles Chamber of Commerce and the *Los Angeles Times*, had led the charge to exert control over the increasingly racialized US-Mexican border to render it impervious to illegal crossings without jeopardizing the flow of cheap and easily controlled farm labor. The border-crossing radio threat from Cuba in the mid-1940s thereby engaged that preexisting defensive and racially tinged worldview.[30]

At the same time, the charge that disruptive NARBA-revising Washington officials gravely threatened their economic livelihood spoke to broader signifiers of a regional identity that resonated among Far Westerners as whole. The 1946 NARBA had, in fact, incited long-standing regional feelings of economic and political vulnerability. The agreement did so by inciting the region's historically noninterventionist and unilateralist leanings in the context of longstanding rural-urban antagonisms. However, to underscore the legitimacy of their grievances, they also leaned on a shared sense of American national identity and belonging to make their case. Disregard of their concerns, this line of argument went, was an affront to their own status and entitlements as Americans. Moreover, in posing a threat to southwestern agriculture, the NARBA revision threatened the national food supply and was therefore a matter of national significance. Taken together, immediate local concerns prompted KFI's advocates to call on the regionally based signifiers of Far Western rural identity to frame the 1946 NARBA as an affront to their specific interests, as well as to larger American national interests. KFI's advocates deployed different layers of their personal identity in a way that privileged the most relevant and useful components of each to make sense of and confront the challenges immediately at hand.[31]

Protests against the 1946 NARBA were not solely limited to the Far West. KFI's defenders also found allies outside the region among the representatives for other clear channel stations, such as WHO of Des Moines, WLS of Chicago, WSM of Nashville, WOAI of San Antonio, and WLW of Atlanta.[32] The Far West also had anti-NARBA allies in other rural sectors of the country. In fact, to torpedo the agreement, the CCBS also sought the support of powerful Farm Bureau affiliates across the country. Consequently, other state- and county-level Farm Bureau organizations, such as those in Texas, Tennessee, Kentucky, Michigan, Georgia, and Illinois, all went on record insisting that the preferences and protections provided to American clear channel stations remain undiminished or ideally enhanced because of their significance to farmers. The Texas organization and its members sounded very much like their California counterparts in claiming that the 1946 NARBA posed a grave risk to their state's citrus crop by threatening the reception of weather reports delivered by clear channel stations to farmers and growers. The Tennessee organization did too, except that it was the state's tobacco rather than citrus growers that the 1946 NARBA allegedly put at risk.[33]

This position, however, was not the consensus view across rural America and not even the consensus within the Farm Bureau. For their own reasons, other regional agricultural lobbying groups were not motivated to oppose NARBA. These groups instead found the clear channel system the source of, rather than the solution to, their grievances. Farm Bureau organizations in Iowa, Indiana, North Carolina, and Ohio broke with their national and other state-level organizations by not championing the sacrosanct status of clear channel preferences. These groups argued that the stations best equipped to serve the needs of their constituencies had been effectively stifled by many of the same clear channel interests now railing against NARBA. The Iowa Farm Bureau lamented that years earlier the station best equipped to serve its members' agricultural interests—WOI of the Iowa State College of Agriculture and Mechanical Arts—had its broadcasting time and power severely limited to accommodate none other than KFI. KFI, a distant California station whose programming did not serve the interests of Iowans, had successfully lobbied the FCC to reject WOI's request to increase its morning broadcasting hours. Those past events cemented the Iowa group's ambivalence to KFI's concerns. Michigan State College of Agriculture and Applied Science, which operated WKAR, and other Michigan broadcasters also broke ranks with their own state's anti-NARBA Farm Bureau and argued against maintaining the current clear channel preferences because the clear channel system did not adequately serve their constituencies.[34]

In short, the fight surrounding the 1946 NARBA is not one easily caricatured as reflective of some simplistic rural-versus-urban dichotomy. Those mobilizing in defense of KFI represented one constituency in a larger domestic battle over what preferences the station licensing system should privilege. This fight had endured in one form or another since radio's earliest days. KFI's advocates claimed that the clear channel system was essential to providing adequate agricultural broadcasting service. That assertion, however, was also contested in other parts of rural America. The claim that KFI was one of the greatest American achievements in rural broadcasting with nationwide significance but whose viability was threatened by the 1946 NARBA was, in fact, locally and regionally conceived. To make their case and fight the challenges the 1946 NARBA revision purportedly presented to their livelihood, KFI's defenders called on deeply rooted signifiers of a regionally based Far Western identity that complemented their broader sense of American identity.

Ultimately, the fight to block the 1946 NARBA for the benefit of KFI was all for naught. The prevailing judgment was that the 1946 NARBA protected more interests than it sacrificed. The smaller local broadcasters, regional broadcasters, and the National Association of Broadcasters concurred with this view, as did the national networks themselves (who feared how their smaller affiliates might be adversely affected by the collapse of the NARBA system). Paul Spearman, general counsel of the Regional Broadcasters Committee, was particularly strong in his support. "I am convinced that this is a step in the right direction," Spearman assured American broadcasters and listeners. "No regional station will suffer as a result of the accords reached." This constellation of interests proved more influential than the sky-is-falling rhetoric deployed by KFI's advocates. The 1946 NARBA interim agreement went into effect as planned on March 29, 1946.[35]

* * *

The 1946 NARBA did not cripple KFI. In fairness, warnings that the 1946 NARBA had insufficient technical requirements to fully protect KFI from interference proved prescient to some degree. Havana's CMQ became a primary source of KFI's interference, just as had been forewarned. However, testing confirmed that CMQ was, in fact, operating well within the technical parameters established by the 1946 agreement; those parameters were supposed to have prevented the very interference that ultimately occurred. It baffled the engineers. Nor was it just CMQ interfering with KFI. A handful of other US and Cuban stations reported interference problems following the 1946 agreement's enactment. That said, the instances of ill-understood interference appeared to ebb and flow, yet never rose to the level of provoking the type of sustained public complaints and dramatic claims of impending doom evidenced in early 1946. In fact, KFI's southwestern listeners had remained remarkably protected from interference in the years that followed; to the extent that border-crossing interference remained a problem after 1946, it was one borne mostly by listeners based in the southeastern United States, due to their proximity to Cuba and other Caribbean-based stations.[36]

And yet fears of an ominous foreign threat that promised to thoroughly drown out US broadcasting with foreign radio intrusions, particularly from nearby Cuba, never fully abated. Familiar gloom-and-doom warnings about foreign radio threats returned in 1948 and 1950. In 1948 the trigger was a bill moving through Congress prohibiting US stations from broadcasting over 50kw of power, prompting warnings that more power might be needed to drown out offending foreign stations, especially those from Cuba. In 1950, it was another NARBA revision that offered additional concessions to Cuba. Redlands grower Arthur Isham, a prominent voice in the protests against the 1946 revision, traveled to Washington to testify against the 1948 bill. He wanted KFI to have more power to drown out any Cuban "rumba" that might be heard. "The air belongs to the United States [and] the farmer is entitled to it," Isham insisted. WGN's treasurer and one of its directors Frank Schreiber concurred, testifying that "if we do not protect our radio programming in this country, we are going to be flooded with programming from foreign countries." However, when pressed to support his claim that the 1946 NARBA was responsible for this vulnerability, Schreiber could only vaguely recall one incident that may have occurred "a year or two ago." When the 1950 NARBA revision was under consideration, this time requiring formal ratification, the Associated Press ran a story fearing for the future of "Farmer Brown." The term was intended to be eponymous for American farmers, who would no longer be able to access the weather, market, and entertainment broadcasts so necessary to their lives. Domestic opposition did, in fact, stall ratification until 1960, but the United States nonetheless adhered to its terms during the intervening decade. And Cuban rumba and other foreign language programs did not overrun the US airwaves in that interim.[37]

Just as the earlier radio conflicts with Panama and Mexico demonstrated, the 1946 fight over NARBA underscored the limits of American power. Because of the nature of radio technology and vulnerability to static interference, a presumably weaker neighbor like Cuba could wreak havoc on the well-developed commercialized American broadcasting system just by broadcasting on frequencies otherwise used by the United States. Cuba's calculations and determination to challenge the United States was a manifestation of more deeply rooted US-Cuban tensions that extended far beyond radio and reflected nearly a half-century of US domination and quasi-colonial control over the island nation. To achieve the specific goal of rolling back the suffocating American radio presence, Cuba followed the same script that Mexico and Panama used: using its proximity to and threatening to wreak havoc on American radio. Doing so came at little cost to itself, while allowing officials to reap the domestic political benefits of standing strong against the American colossus. In fact, following Cuba's successful 1959 communist revolution that freed it from the US imperial orbit (save for Guantanamo Bay), radio wars and the efforts to deliberately project unwelcome programming into each country became and remained a standard facet of the interactions between the two subsequently hostile nations.[38] Meanwhile, the results of the 1946 NARBA fight (and again in its 1950 incarnation) also followed the Mexican and Panamanian precedents: the United States ultimately agreed to concessions once fiercely resisted in the face of vociferous and enduring objections to a perceived surrender.

As Americans grappled with the limits to their power as it pertained to radio, the fight against the 1946 NARBA inflamed the divisions within the American polity. Those divisions reflected the enduring strength of local and regional identities in the United States. The effort to resist perceived Cuban encroachments inspired NARBA protesters from California and Arizona to seize on elements of their local, regional, and national identities to demand protection for certain rights and interests to which they felt entitled. As Joseph Fry notes in his essay on the intersection of regionalism and US foreign policy, "place matters in how Americans have responded to and influenced the formation and implementation of US foreign policy."[39] Although the southwestern farmers and growers of this account may have failed to influence the trajectory of US foreign policy as it pertained to NARBA, the enduring localism and regionalism embedded in their outlooks shaped how they responded.

It is through exploring their responses that we encounter a story of larger national significance in American history. The United States was (and is) a nation where regional identity had been historically valued, but also one where long festering rural-urban tensions had left many Americans in the Southwest feeling dependent, vulnerable, and disregarded by the increasingly dominant urban sector. The 1946 NARBA provided the vehicle for many of those southwesterners to air fundamental grievances rooted in a regionally based Far Western identity. They lamented NARBA's alleged neglect of rural interests and, in the process, expressed more deeply rooted antagonisms toward the unwelcome disruptions instigated by a distant and seemingly unresponsive federal government. Even though the worst fears never materialized, the failure to block the 1946 NARBA agreement only confirmed an enduring and still palpable sense of divide between the farmers and growers of California and Arizona versus an imprudent, if not outright abusive, urban-oriented, and internationalist-minded federal government.

CONCLUSION

From "Whistling and Singing 'La Paloma'" to "No Way, José"

A Century of Continuity and Change in Communications, Identity, and Borders

The stories told in the preceding chapters underscore how various aspects of international connectivity had a discernible impact on national broadcasting policies, practices, and cultures. The enduring historical significance of the global movement of peoples loomed large in the controversies that unfolded. American listener opposition to broadcasts in languages other than English by the 1930s reflected a reaction, at least in part, to the high levels of immigration the United States experienced in the late nineteenth and early twentieth centuries in the context of a looming world war. The late 1930s conflict over whether to use Cajun or standard French in LSU's educational broadcasts cannot be disentangled from the enduring influences and idealization of the eighteenth-century French-Canadian migration to Louisiana. That American Walter Lemmon and his cohort aspired to use radio to teach English to predominantly Spanish-speaking Latin America was itself a manifestation of the enduring political and linguistic consequences of European colonization and wars in the Western Hemisphere that dated back to the fifteenth century. By the twentieth century, the United States was itself a well-established imperialist and colonialist power with far-flung interests and a dominating presence across much of Latin America and the Caribbean. In the context of this book, controversial American radio activities and policies in and toward Mexico, Panama, and Cuba worked in tandem with dismissive attitudes toward those nations' radio priorities, which helped fuel the broader anti-Americanism that arose in conjunction with the US imperialism of that era. Popular and official reactions to American radio excesses from those three nations ultimately had discernible effects on ideas about and practices embedded in those nations' then-emergent national broadcasting systems.

Through those explorations, this book examines dubiously based convictions rooted in misunderstandings of radio's power. Misplaced beliefs in radio's singular ability to propel change and impose its influence distorted approaches and reactions to radio programming and radio policies. Protests against foreign language radio programming heard on American radio in the 1930s and early 1940s offers one such example. Hyperbolic fears of a radio-amplified "fifth column" of purportedly disloyal foreigners living amidst Americans were not commensurate with any real threat. Reactions against this perceived foreign radio menace instead served to underscore the English language's enduring centrality to American national identity. The linguistic purism that infused LSU's contemporaneous French Radio Project alongside a distorted nostalgia for Louisiana's supposedly purer French traditions rendered that initiative singularly ill-suited to engaging, much less educating, its target Cajun audience. Or at least that would have been the case if it had not quickly collapsed under the weight of fieldworker Louise Olivier's disinterest in the radio project's original educational premise. The comparably flawed "Basic English" initiative pursued during those same years also had little chance of engaging and educating a vast audience in Latin America and the Caribbean given its inattentiveness to the full range of complexities that can encourage or discourage listener engagement.

The foreign policy–focused chapters of the book underscore questionable convictions about the power of radio to serve and achieve distinctly American goals regardless of context. Up until the eve of American entry into World War II, a combination of American officials' willful disregard of and outright blindness to their foreign counterparts' perspectives infused attitudes and actions surrounding radio policy. Counterproductive dismissiveness toward Mexico's radio priorities sustained a border radio dispute for more than a decade. A comparable disregard for Panama's radio priorities plagued US-Panamanian radio relations across two world wars. Only the emergence of larger perceived threats to hemispheric security during the Second World War in the face of escalating anti-American animosities in Mexico and Panama finally compelled the United States to change course and agree to concessions. In 1946, US officials appeared to learn from past mistakes by quickly agreeing to Cuba's demands for revisions to the 1941 radio treaty that resolved the earlier radio dispute with Mexico. In so doing, however, those same officials antagonized deeply rooted US domestic divisions. Far Western agricultural interests attempted to block the pact by claiming federal authorities, long distrusted in the Far West, had unjustifiably appeased undeserving Cubans at the expense of their rights and interests as Americans. In each example explored in the preceding pages, radio did not yield the unity so many predicted would inevitably follow its proliferation. Retired naval commander George Sweet obviously missed the mark with his 1923 forecast that Americans "would soon be whistling and singing 'La Paloma'" thanks to radio-driven ties with the Spanish-speaking peoples of Latin America.[1]

In the varied cases explored here, radio's aural attributes loomed large. Perceptions of radio threats often captured how the act of listening triggered cognitive processes of imagination. In the absence of visual cues, listeners imagined the people and their surroundings based on the audio they heard. When such content provoked perceptions of a foreign threat, those visualizations often included the national territory that the otherwise invisible radio waves had purportedly violated. The rigor of this mental imaging exercise imbued such visualizations with power and emotion for their creators. A listener's existent body of knowledge and assumptions specific to their historical and situational positioning drove this imaginative process that produced pictures in the mind's eye. This visualization process helps explain why some programs and initiatives provoked such vitriolic reactions; listeners' perspectives and preexisting convictions that infused such visualizations were not necessarily in sync with particular broadcasters' or policymakers' objectives.

In the years since 1946, there have been some indications of changing patterns and approaches to international communications and media that were seemingly better suited to achieving the desired objectives. There was, for example, the Cold War era's Voice of America (VOA), an official US international radio initiative that sought to build pro-US sentiment in antagonistic Communist bloc countries. It appeared to successfully engage a broad swath of listeners living in the manipulated media markets of the Communist world. With its broadcasts heard in hostile territory despite the prohibitions of the respective authorities, this effectiveness is often credited to VOA's skill in presenting credible information that exposed the deceits of various state-controlled media outlets in ways that proved convincing to many in the target audiences. By the 1980s, as the communist systems in Eastern Europe began to unravel, VOA—the more successful and enduring international radio progeny of the World Wide Broadcasting Foundation—had more than 100 million regular listeners worldwide who, by choosing to listen often in defiance of their nation's laws, seemingly signaled their appreciation of its programming quality.[2]

Such outcomes also appear evident beyond radio. Since the last decades of the twentieth century, American movies evidently enjoyed growing popularity throughout much of the world. The US prioritization of free trade policies since World War II worked in conjunction with English's status as either the primary or most popular secondary language in what proved to be key foreign markets. That combination helped pry open many such markets to American motion pictures. By 2015 US-produced films had a 29 percent share of total global box office receipts, outpacing European films by more than 8 percent and Chinese-produced films by more than 20 percent. These statistics suggest that American films had a cross-cultural resonance capable of attracting expansive international audiences.[3]

From the 1960s through the 1990s, satellites followed by high-capacity fiber optic cables further transformed the international communications infrastructure. Content and information could move ever more quickly across the globe. The precision with which transmissions could be directed from one point to another on opposite sides of the world far exceeded the capabilities and reliability of shortwave radio broadcasting. The capacity to transmit video content throughout the planet was unprecedented. These evidently improving capabilities are what inspired US President Lyndon Johnson's optimistic prediction that communications satellites would foster a "partnership" among "all nations under the sun and stars."[4]

Once again, as was the case with the predictions affixed to radio and so many other communications technologies before it, this utopian-like optimism was misplaced. As a case in point, VOA's perceived postwar successes in influencing hearts and minds can easily be overstated. In the People's Republic of China, for example, the Voice of America did indeed by the 1960s build a significant audience in the face of China's credibility-challenged official news sources. However, widespread doubts among the Chinese about their official media's accuracy did not automatically imbue VOA with unquestioned credibility of its own. VOA's position as a well-known official US government media outlet in the context of Sino-American enmity raised substantive doubts among Chinese listeners about VOA's reliability. VOA, in fact, was just one of several sources of perceived dubious credibility many Chinese listeners turned to in conjunction with official state media. Chinese listeners often attempted to cobble together a more accurate picture of the present from a composite of information gleaned from individual sources not singularly trusted as wholly accurate sources by themselves. It is an excellent example of audience agency at work. Ultimately, the authoritarian regime that the VOA hoped its programming could help topple is still firmly in power and Sino-American relations have remained fraught into the present.[5]

Internationally popular Hollywood films also fell short of serving as a singular global unifying force cultivating common ground among diverse peoples. Just as John Brinkley's radio exploits from Mexico simultaneously inspired both passionate opposition to and support of his presence on American airwaves,

Hollywood films could also provoke comparably contradictory reactions from international audiences. When, for example, the 1993 blockbuster *Jurassic Park* opened to enthusiastic audiences in hundreds of theaters across France and Spain, it concurrently sparked in those two countries widespread vocal and visible protests railing against the movie's pervasive presences in theaters across both nations. Those protesting objected to the limited number of screens left available to show purportedly superior home-grown productions.[6]

In other instances, a movie that is popular in multiple international markets is not necessarily engaging the same cultural sensitivities of those different audiences. A year before *Jurassic Park*, for example, *Like Water for Chocolate* was celebrated as the first Spanish-language Mexican film to earn widespread popularity with an American audience. However, different versions of the film were released in the United States and Mexico to account for different audience interests and sensitivities. In the US version, producers dubbed scenes for the purpose of changing accents to ones deemed more palatable for an American audience. More recently, China and various governments in the Arabic world have been particularly insistent on expunging LGBTQ representation from American films. In other words, the same nominal film being shown in multiple countries can be deliberately altered so that it does not engage the same range of cultural sensitives of its varied and diverse audiences across different parts of the world.[7]

Cross-border media flows and their potential influence are also still restricted by present-day treaty arrangements, just as NARBA did in the 1940s. Counterintuitively, the North American Free Trade Agreement (NAFTA) between the United States, Canada, and Mexico, with its "cultural exemption clause," is one such accord. That clause, first included in the bilateral 1987 US-Canada Free Trade Agreement, was retained in the 1993 NAFTA. It allows for restrictions to be placed on the cross-border flow of "cultural goods," whose products were defined to include traditional radio and television broadcasts. In its most recent incarnation, the 2020 NAFTA revision acknowledged that "states have a sovereign right to preserve, develop and implement their cultural policies, to support their cultural industries for the purpose of strengthening the diversity of cultural expressions, and *to preserve their cultural identity.*" Neither Mexico's negotiators nor those for Canada's French-speaking Quebecois minority pushed for the clause's continued inclusion. Past experiences indicated their respective constituencies had limited interest in English language media content imported from the United States. Instead, representatives for English-speaking Canada fiercely insisted the clause be retained. They, more than their counterparts, fretted over opening the floodgates to cross-border American media influence. These concerns were not unfounded. The United

States had become an increasingly dominant exporter of its media and other cultural products during the latter half of the twentieth century. Its products—particularly films—did indeed appear to have great appeal to culturally similar English-speaking Canadians, far beyond what was evidenced among their French-speaking compatriots. The United States ultimately abandoned its efforts to exclude the clause from the 2020 NAFTA revision when Canada threatened to scuttle the entire agreement if the exemption was not retained.[8]

Despite such examples, optimistic predictions of a world destined to be made closer by ever-improving communications endure. "The spread of the Internet will enhance global connectivity that fosters more planetary relationships and less ignorance," was one such hopeful expectation cited by the Pew Research Center in a 2014 assessment of expert predictions.[9] In fact, in a media world that offers abundant and varied content, it is easy to ignore or be oblivious to purportedly disagreeable content. In the 1930s, physically moving the radio tuner across the dial from one station to another—made possible by Walter Lemmon's invention of "single dial tuning"—increased the likelihood of encountering programming one might find displeasing. In the twenty-first century, a remote control, computer mouse, or handheld device facilitates immediate navigation to preferred content among seemingly countless options. The algorithms used by streaming and social media providers, which recommend new content based on past choices, further confine media consumers to a "bubble" or "echo chamber." To be sure, the earlier generations of media consumers explored in the preceding chapters demonstrated propensities to seek out the content they personally preferred while avoiding content they disliked, and objecting to it when avoidance failed. These newer tools, however, can effectively enhance a media consumer's success in capitalizing on those time-tested instincts and thereby remain better insulated in a preferred bubble. It is a dynamic that has helped inflame an increasingly toxic and polarized American political and social environment when competing and contradictory bubbles collide, not just amidst divisive partisan political debates and campaigns but across broader social media exchanges.[10]

In this context, the dynamics of global interconnectedness continue to fuel perceived threats to the nation emanating from foreign languages, foreign peoples, and perceptions of difference. In spring 2022, Alabama Republican governor Kay Ivey was running for reelection and released a campaign ad displaying images of supposedly undocumented immigrants from Latin America purportedly streaming across the US-Mexican border and eventually into Alabama, allegedly at the behest of President Joe Biden, a Democrat. Ivey assured her constituents that she took the threat seriously. Though Alabama does not share a border with Mexico, she mobilized her state's National Guard troops for deployment in Texas in cooperation with a larger "Border Strike Force" organized by other Republican governors. If the flow into Alabama of these "illegals" did not abate soon, she warned, "we're all going to have to learn Spanish." That warning sets up Ivey to deliver her punchline: "my message to Biden: no way, José!" The controversial ad generated nationwide media attention, both positive and negative. She ultimately won reelection in a landslide that autumn, earning more than twice the number of votes secured by her challenger.[11]

To the extent this ad helped effectively mobilize her constituencies, its message relied upon the historical racialization of the Spanish language and the Latin Americans who spoke it. To the delight of her supporters, she unapologetically dismissed allegations from political opponents and national critics that her "No Way, José" ad, as it came to be called, trafficked in racism. "There is nothing racist with telling the truth about the disaster Joe Biden is causing with illegals invading our country." The ad was, to be fair, a far cry from the abject racism that infused attitudes and policies toward immigration in the 1920s. It did, however, harness the same enduring linguistic intolerance that more than 260 years earlier provoked Benjamin Franklin's fears, as noted in chapter 1, that German settlement in Pennsylvania threatened to turn the then-British colony into a German one and force its English-speaking residents to "live as in a foreign country." More to the point, the deliberately prioritized focus on spoken language juxtaposed with visuals of darker-skinned Latin American migrants is also a historically potent form of "othering" in which Spanish-speaking Latin Americans, as well as native-born multilingual Americans of Latin American descent, have long been cast as outsiders and potential threats to the nation. "Here in Alabama," Ivey assured her audience, "we're gonna enforce the law." The thirty-second video, in fact, avoided any specific reference to the substantive legal issues at play. Many meaningful debates about border security often pointed to cross-border drug and human trafficking or the challenges local government infrastructures faced in handling large numbers of migrants and refugees. Such substance was, in fact, superfluous to this ad. It was far more efficient and effective to use its half-minute to channel the deeply rooted English-centric nativist liberalism that still resonates in many quarters of American political culture. This nativist liberalism, communications scholar Hector Amaya argues, presumes that the safety, security, and longevity of the American nation requires monolingualism based on the enduring primacy of English and immigrant assimilation on that basis.[12]

Ivey's media message thus tapped into persisting anxieties about language and perceived threats to identity and society that transcend historical moments and peoples. In the 1930s and 1940s, aggrieved American listeners often underscored the role of language in the perceived threats manifested by both domestic and foreign broadcasting. They lodged protests against domestically based foreign language broadcasts and the foreign-based broadcasts from places like Mexico and Cuba. The nation itself was not always the perceived target of such audibly manifested threats. Louise Olivier identified a looming menace in the form of spoken (and racialized) Cajun; allowing it on the radio threatened to further degrade and ultimately erase Louisiana's traditional and supposedly superior French heritage specific to that state's history. When Cuba received rights to broadcast on the same frequency as US-based KFI, Far Western farmers who relied on the station objected that their rights as Americans had been abridged by malfeasant federal authorities. However, the ways in which they articulated the perceived threats to their livelihood from radio broadcasts of Cuban "rumba" underscored a distinct Far Western identity that distinguished them from other regional American identities.

Nor were these instincts specific to Americans. In the 1930s, Mexican authorities confronted domestic political pressures after authorizing the operation on Mexican soil of so-called border blasters by English-speaking "Yankees" such as John Brinkley. For many Panamanians, US domination of radio during Panama's earliest decades of existence became emblematic of the United States' multilayered colonization and subordination of their country. Cuba's determination to revise the 1946 NARBA in its favor was in direct response to the inundation of its airwaves with US-based radio programming to the point that American radio had stunted the development of Cuban broadcasting. Perhaps the absence of fierce pushback to the short-lived effort to teach Basic English over the radio to Latin America and the Caribbean is an exception to these patterns. If so, it is an exception only because too few people ever heard the broadcasts for them to have any meaningful long-term impact.

Past or present, there is nothing especially novel about media consumers being drawn to content that speaks to and amplifies existing beliefs and convictions. There is nothing unusual about users who seek to avoid content that contradicts or complicates those existing beliefs, and objecting when such content intrudes into their world. These patterns and reactions have paradoxically long coexisted alongside the time-worn proclamations celebrating the potential of new communications to bind peoples together. Such enduring claims notwithstanding, when content subjectively determined to be threatening is encountered, the result is often all too predictable: vociferous complaints and efforts to marginalize both the offending message and the peoples it represents. This dynamic was part of the story of early radio and remains so in more recent media environments. Such reactions, then and now, are often rooted in the notions of identity, nationalism, and the enduring prejudices that have been and remain prominent in our world.

Notes

Introduction

1. B. Q. Morgan (Professor of German, University of Wisconsin) and John Van Horne (Professor of Romance Languages, University of Illinois), "Bibliography of Modern Language Methodology for 1924," in *Modern Language Journal* 9.8 (1925), 495; Martin P. Rice, "The Future of Radio Broadcasting," in *The Monogram* 1.4 (1924), 7, General Electric Papers, Schenectady Museum Archives, Schenectady, New York. "Talkies" quote, c. 1920s, from untitled and undated essay, file 485a, Box 198, Series 14 (General History), George H. Clark Radioana Collection, Archives Center at the National Museum of American History at the Smithsonian Institution.

2. C. H. Mercer, "Radio and Aural Comprehension," *Modern Language Journal* 15.5 (1931), 328, 336. E. F. Engel, "Radio as a Medium of Modern Language Instruction," *Modern Language Journal* 20.3 (1935), 171. Katherine D. Blake, "The New Outlook," *Journal of Education* 96.3 (1922), 75.

3. Sweet to Hughes, 27 March 1923, 812.74/189, Department of State Central Decimal Files 1910–29, Record Group 59, United States National Archives at College Park, Maryland (hereafter abbreviated as DSNA, followed by the years covered in the collection cited, e.g., 1910–29, 1930–39, etc.). Opening address by J. G. Harbord, 1927 International Radiotelegraph Conference, 15 October 1927, folder "Radio: Conferences-International-Washington (1927)—Addresses by Hoover, Coolidge, et al., 1927, October and November," Box 492, Herbert Hoover Commerce Department Papers, Herbert Hoover Presidential Library, West Branch, Iowa. Castle to Professor William F. Ogburn, University of Chicago, 16 April 1931, 800.76/29, Box 4360, DSNA 1930–39. The anonymous musing is from the same unattributed and undated piece from the Smithsonian Institution's Clark Radioana cited in n. 1 above. See also Simon Potter,

Wireless Internationalism and Distant Listening: Britain, Propaganda, and the Invention of Global Radio, 1920–1939 (New York: Oxford University Press, 2020). For background information on Sweet, see Lillian C. White, *Pioneer and Patriot: George Cook Sweet, Commander, U.S.N., 1877–1953: A Biography* (Delray Beach, FL: The Southern Publishing Company, 1963), 103, 107; Hugh G. J. Aitken, *The Continuous Wave: Technology and American Radio, 1900–1930* (Princeton, NJ: Princeton University Press, 1985), 289–90.

4. Jonathan Rosa and Nelson Flores, "Unsettling Race and Language: Toward a Raciolinguistic Perspective," *Language in Society* 46.5 (November 2017), 621–47; Anne H. Charity Hudley, "Language and Racialization," in Ofelia García, Nelson Flores, and Massimiliano Spotti, eds., *The Oxford Handbook of Language and Society* (New York: Oxford University Press, 2017), 381–402; Andrew D. Wong, Hsi-Yao Su, and Mie Hiramoto, "Complicating Raciolinguistics: Language, Chineseness, and the Sinophone," *Language and Communication* 76 (2021), 131–35; Jillian R. Cavanaugh, "Accent Matters: Material Consequences of Sounding Local in Northern Italy," *Language and Communications* 25.2 (2005), 127–48; Dan Worth, "As a Teacher, Does Your Accent Matter?," *Times Educational Supplement*, 13 September 2019, https://www.tes.com/magazine/archived/teacher-does-your-accent-matter (this and all subsequent links cited in the notes were working links as of 23 December 2023).

5. Menahem Blondheim, *News over the Wires: The Telegraph and the Flow of Public Information in America, 1844–1897* (Cambridge, MA: Harvard University Press, 1994), 191.

6. David Paull Nickles, *Under the Wire: How the Telegraph Changed Diplomacy* (Cambridge, MA: Harvard University Press, 2009); John A. Britton, "'The Confusion Provoked by Instantaneous Discussion': The New International Communications Network and the Chilean Crisis of 1891–1892 in the United States," *Technology and Culture* 48.4 (2007), 729–57.

7. Stephen Kern, *The Culture of Time and Space, 1880–1918* (Cambridge, MA: Harvard University Press, 1983), 259–86; Lyndon Johnson is quoted in James Schwoch, *Global TV: New Media and the Cold War, 1946–1969* (Urbana: University of Illinois Press, 2009), 152. Hugh R. Slotten, *Beyond Sputnik and the Space Race: The Origins of Global Satellite Communications* (Baltimore: Johns Hopkins University Press, 2022); Tom McCarthy, "How Russia Used Social Media to Divide Americans," *The Guardian*, 14 October 2017, https://www.theguardian.com/us-news/2017/oct/14/russia-us-politics-social-media-facebook; Scott W. Harold, Nathan Beauchamp-Mustafaga, and Jeffrey W. Hornung, "Chinese Disinformation Efforts on Social Media," Rand Corporation, 2021, https://www.rand.org/pubs/research_reports/RR4373z3.html.

8. Thomas Bender, *A Nation Among Nations: America's Place in World History* (New York: Hill and Wang, 2006), 3; Derek W. Vaillant, *Across the Waves: How the United States and France Shaped the International Age of Radio* (Chicago: University of Illinois Press, 2017), 4.

9. Potter, *Wireless Internationalism and Distant Listening*; Suzanne Lommers, *Europe—On Air: Interwar Projects for Radio Broadcasting* (Amsterdam: Amsterdam University Press, 2012); Joelle Neulander, *Programming National Identity: The Culture of Radio in 1930s France* (Baton Rouge: Louisiana State University Press, 2009), 160–84; Andrea Stanton, *"This*

Is Jerusalem Calling": State Radio in Mandate Palestine (Austin: University of Texas Press, 2013); Michael A. Krysko, *American Radio in China: International Encounters with Technology and Communications, 1919–1941* (New York: Palgrave Macmillan, 2011); Daqing Yang, *Technology of Empire: Telecommunications and Japanese Expansion in Asia, 1883–1945* (Cambridge, MA: Harvard University Press, 2010). Daniel Headrick's influential and overarching history of global communications through World War II also emphasized the contrast between widespread expectations of telecommunications fostering "greater harmony between peoples and nations" dating back to the telegraph's emergence and outcomes that deployed international telecommunications as a "weapon in the rivalries of great powers." The entanglement of developing telecommunications networks in fraught international environments, he argues, "contributed to the atmosphere of tension and misunderstanding that was beginning to build up at the turn of the century." See Daniel R. Headrick, *The Invisible Weapon: Telecommunications and International Politics, 1851–1945* (New York: Oxford University Press, 1991), 8–9, 111.

10. On visualization, see Susan J. Douglas, *Listening In: Radio and the American Imagination, from Amos 'n' Andy and Edward R. Murrow to Wolfman Jack and Howard Stern* (New York: Times Books, 1999), 4–8, 11–12, 22–36 (esp. 26–27); see also Edward D. Miller, *Emergency Broadcasting and 1930s American Radio* (Philadelphia: Temple University Press, 2003), 7–10; and John Durham Peters, *Speaking Into the Air: A History of the Idea of Communication* (Chicago: University of Chicago Press, 1999), 214–17. On historical specificity and societal positioning, see Kate Lacey, *Listening Publics: The Politics and Experience of Listening in the Media Age* (Malden, MA: Polity Press, 2013), 24.

11. Nina Sun Eidsheim, "Marian Anderson and 'Sonic Blackness' in American Opera," *American Quarterly* 63.3 (2011), 644–45; Barbara Dianne Savage, *Radio, War, and the Politics of Race, 1938–1948* (Chapel Hill: University of North Carolina Press, 1999), 6–7. The stylistic decision to capitalize "Black" but not "white" reflects the approach adopted by the Associated Press and other major media outlets. See David Bauder, "AP Says It Will Capitalize Black but Not White," 20 July 2020, *AP News*, https://apnews.com/article/entertainment-cultures-race-and-ethnicity-us-news-ap-top-news-7e36c00c5af0436abc09e051261fff1f.

12. Gary Frost, *Early FM Radio: Incremental Technology in Twentieth-Century America* (Baltimore: Johns Hopkins University Press, 2010), 41; Stanford C. Hooper, "Keeping the Stars and Stripes in the Ether," *Radio Broadcast*, June 1922, 125–32.

13. David E. Nye, *Technology Matters: Questions to Live With* (Cambridge, MA: MIT Press, 2006), esp. 15–31, 43–44, 47, 49, 61–64, 105–7, 185–207; see also Merritt Roe Smith and Leo Marx, eds., *Does Technology Drive History? The Dilemma of Technological Determinism* (Cambridge, MA: MIT Press, 1994).

14. Aitken, *The Continuous Wave*; Susan J. Douglas, *Inventing American Broadcasting, 1899–1922* (Baltimore: Johns Hopkins University Press, 1989); Michele Hilmes, *Hollywood and Broadcasting: From Radio to Cable* (Urbana: University of Illinois Press, 1990); Michele Hilmes, *Radio Voices: American Broadcasting, 1922–1952* (Minneapolis: University of Minnesota Press, 1997); Jason Loviglio, *Radio's Intimate Voice: Network Broadcasting and Mass Mediated Democracy* (Baltimore: Johns Hopkins University Press, 2005); Robert

W. McChesney, *Telecommunications, Mass Media, and Democracy: The Battle for Control of U. S. Broadcasting, 1928–1935* (New York: Oxford University Press, 1993); Susan Smulyan, *Selling Radio: The Commercialization of American Broadcasting, 1920–1934* (Washington, DC: Smithsonian Institution Press, 1994); and Hugh R. Slotten, *Radio and Television Regulation: Broadcasting Technology in the United States, 1920–1960* (Baltimore: Johns Hopkins University Press, 2000).

15. Clifford Doerksen, *American Babel: Rogue Radio Broadcasters of the Jazz Age* (Philadelphia: University of Pennsylvania Press, 2005); Ari Y. Kelman, *Station Identification: A Cultural History of Yiddish Radio in the United States* (Berkeley: University of California Press, 2009); Elena Razlogova, *The Listener's Voice: Early Radio and the American Public* (Philadelphia: University of Pennsylvania Press, 2011); Alexander Russo, *Points on the Dial: Golden Age Radio Beyond the Networks* (Durham, NC: Duke University Press, 2010).

16. Vaillant, *Across the Waves*; Michele Hilmes, *Network Nations: A Transnational History of British and American Broadcasting* (New York: Routledge, 2012).

17. Jason Loviglio, a historian of radio at the University of Maryland-Baltimore County, made this comment on 22 October 2011 at the American Studies Association Annual Meeting in Baltimore as part of his participation on a panel entitled "Behind *The Wire*." His presentation was titled "Radio Free Baltimore: Neoliberal Transformation on the Local Public Airwaves." See also Jason Loviglio, "Moments of Danger: The Struggle for Community-based Public Radio in Baltimore," *Journal of Alternative and Community Media* 4.4 (2019), 42.

Chapter 1. "Broadcasting in the Language of the Enemies of Civilization"

1. Earl Andrews to the FRC, 3 December 1931, Box 193, Box 287, File 67–4, Record Group 173, Federal Communications Commission, Office of the Executive Director, General Correspondence, 1927–46, United States National Archives, College Park, Maryland (hereafter FCC Records 1927–46).

2. For early broadcasting's emphasis on more locally attuned and interest-specific programming, including those programs that targeted immigrant and foreign language communities, see Lizabeth Cohen, *Making a New Deal: Industrial Workers in Chicago, 1919–1939* (New York: Cambridge University Press, 1990), 129–42, esp. 133–34; Smulyan, *Selling Radio*, 20–21; Douglas, *Listening In*, 61–62; Nathan Godfried, *WCFL: Chicago's Voice of Labor, 1926–1978* (Urbana: University of Illinois Press, 1997), 113; Robert Hilliard and Michael Keith, *The Quieted Voice: The Rise and Demise of Localism in American Radio* (Carbondale: University of Southern Illinois Press, 2005), 24–27, 33; and Derek W. Vaillant, *Sounds of Reform: Progressivism and Music in Chicago, 1873–1935* (Chapel Hill: University of North Carolina Press, 2003), 236.

3. Douglas, *Listening In*, 74; see also Hilmes, *Radio Voices*, 34–74 and Smulyan, *Selling Radio*, 27. "Last Night the Best Radio Test Week," *New York Times*, 29 November 1924, p. 2.

4. Dolores Inés Casillas, *Sounds of Belonging: U. S. Spanish-Language Radio and Public Advocacy* (New York: New York University Press, 2014), 6, 24–26 (direct quote is from

p. 26). Jessica Gienow-Hecht, *Sound Diplomacy: Music and Emotions in Transatlantic Relations, 1850–1920* (Chicago: University of Chicago Press, 2009), esp. 1, 41.

5. "Consuls Feted by Rotary: Representatives of Eleven Nations Honor Guests at Luncheon; Mexican Music Heard," *Los Angeles Times*, 13 October 1923, p. II-1.

6. Dennis Baron, *The English-Only Question: An Official Language for Americans?* (New Haven: Yale University Press, 1990), 66 (Franklin quoted on this page), 87–88; Hector Amaya, "Nativist Liberalism and the Disciplining of Spanish Language Media," Marwan M. Kraidy, ed., *Communication and Power in the Global Era* (New York: Taylor and Francis, 2013), 14–15, 25.

7. Matthew Frye Jacobson, *Barbarian Virtues: The United States Encounters Foreign Peoples at Home and Abroad, 1876–1917* (New York: Hill and Wang, 2000), 61; Matthew Frye Jacobson, *Whiteness of a Different Color: European Immigrants and the Alchemy of Race* (Cambridge, MA: Harvard University Press, 1998), 78–92.

8. Douglas C. Baynton, *Defectives in the Land: Disability and Immigration in the Age of Eugenics* (Chicago: University of Chicago Press, 2016), esp. 1–10.

9. Baynton, *Defectives in the Land*, esp. 2, 4, 133–34; for one particular state-level context, see Heather L. McCrea, "Treating Delinquent and Feebleminded Juveniles at the Beloit Industrial School for Girls in Early Twentieth-Century Kansas," *Journal of the History of Medicine and Allied Sciences*, 79.3 (2024), 193–211. See also Gary Gerstle, *American Crucible: Race and Nation in the Twentieth Century* (Princeton, NJ: Princeton University Press, 2001), 80–127, esp. 82; Thomas Paul Bonfiglio, *Race and the Rise of the Standard American* (New York: De Gruyter, 2002), esp. 136–37; Terrence G. Wiley, "Continuity and Change in the Function of Language Ideologies in the United States," Thomas Ricento, ed., *Ideology, Politics, and Language Policies: Focus on English* (Philadelphia: John Benjamins Publishing Company, 2000), 82–83; Tony Judt and Denis Lacorne, "The Politics of Language," Judt and Lacorne, eds., *Language, Nation, and State: Identity Politics in a Multilingual Age* (New York: Palgrave MacMillan, 2004), 9–11.

10. Charles M. Adams, "Radio and Our Spoken Language: Local Differences Are Negligible, But Radio Shows Up Personalities," *Radio News*, September 1927, 208, 293 (direct quotes are from p. 208).

11. Douglas, *Listening In*, 56; McChesney, *Telecommunications, Mass Media, and Democracy*, 14–19; Smulyan, *Selling Radio*, 20, 22. Douglas B. Craig, *Fireside Politics: Radio and Political Culture in the United States, 1920–1940* (Baltimore: Johns Hopkins University Press, 2000), 12. "Annual Report of the Federal Radio Commission to the Congress of the United States for the Fiscal Year Ended June 30, 1927" (Washington DC: US Government Printing Office, 1927), 8, accessed at https://docs.fcc.gov/public/attachments/DOC-338385A1.pdf.

12. United States Department of Commerce, Bureau of the Census, *Historical Statistics of the United States, Colonial Times to 1970*, part 2 (Washington, DC: US Government Printing Office, 1975), 796, accessed online at https://www2.census.gov/library/publications/1975/compendia/hist_stats_colonial-1970/hist_stats_colonial-1970p2-chR.pdf; "Annual Report of the Federal Radio Commission 1927," 2–3; Craig, *Fireside*

Politics, 12; Jesse Walker, *Rebels on the Air: An Alternative History of Radio in America* (New York: NYU Press, 2001), 34–35; Douglas, *Listening In*, 56, 62; McChesney, *Telecommunications, Mass Media, and Democracy*, esp. 18–29, 260–62; Smulyan, *Selling Radio*, esp. 8–9, 42–44, 62–64, 166–68; Fritz Messere, "The Davis Amendments and the Federal Radio Act of 1927: External Pressures in Policy Making," J. Emmett Winn and Susan L. Brinson, eds., *Transmitting the Past: Historical and Cultural Perspectives on Broadcasting* (Tuscaloosa: University of Alabama Press, 2005), esp. 53, 55, 57; Josh Shepperd, "Infrastructure in the Air: The Office of Education and the Development of Public Broadcasting in the United States," *Critical Studies in Media Communications* 31.3 (2014), 234–35; Razlogova, *The Listener's Voice*, 4; Russo, *Points on the Dial*, 4.

13. McChesney, *Telecommunications, Mass Media, and Democracy*, 29; Craig, *Fireside Politics*, 12.

14. McChesney, *Telecommunications, Mass Media, and Democracy*, esp. 18–29.

15. Douglas, *Listening In*, 106–7, 113, 122; Hilmes, *Radio Voices*, xix, 1–33, 75–96; Vaillant, *Sounds of Reform*, 234–69; Smulyan, *Selling Radio*, esp. 9.

16. For the marginalization of smaller broadcasters, see McChesney, *Telecommunications, Mass Media, and Democracy*, 27–29, 63–91; Slotten, *Radio and Television Regulation*, 58. For the FRC's first survey of foreign language broadcasters, see "Sixth Annual Report—Federal Communications Commission for the fiscal year ended June 30 1940," 52, https://docs.fcc.gov/public/attachments/DOC-308656A1.pdf. See also Gerd Horten, *Radio Goes to War: The Cultural Politics of Propaganda during World War II* (Berkeley: University of California Press, 2002), 70; E. F. Engel, "The Broadcasting of Modern Foreign Languages in the United States: Second Survey," *Modern Language Journal* 22.8 (1938), 626; "Alien Programs on 199 Stations," *New York Times*, 7 December 1940, p. 8; "News Notes on Radio: National Defense Keynotes the Activity of Commissions in Washington," *New York Times*, 22 December 1940, p. 110. For the transmission power level of network stations, see McChesney, *Telecommunications, Mass Media, and Democracy*, 29, 110.

17. Godfried, *WCFL*, 71–105, 135–65; Nathan Godfried, "Struggling over Politics and Culture: Organized Labor and Radio Station WEVD during the 1930s," *Labor History* 42.4 (2001), 347–69, esp. 357; Craig, *Fireside Politics*, 68, 74; McChesney, *Telecommunications, Mass Media, and Democracy*, 63–72; Vaillant, *Sounds of Reform*, 264–67 (direct quote from p. 265). "Four Radio Stations in Brooklyn Face End," *New York Times*, 2 March 1934, p. 16.

18. Engel, "The Broadcasting of Modern Foreign Languages in the United States," 626; "US and Canadian Stations by Frequency," *Radio World*, 13 December 1930, p. 28; "North American Radio Log," *Radio Guide*, 8 January 1939, p. 18.

19. "Invader Holds Up Studio, Broadcasts Hate for Nazis," *Boston Globe*, 28 December 1939, pp. 1, 9; "Seizes Station to Protest War," 29 December 1939, p. 6 (contains Bielecki's quote threatening to shoot anyone who moved); "Pole Steals Air!," *Radio-Craft*, April 1940, p. 584.

20. "Invader Holds Up Studio, Broadcasts Hate for Nazis," *Boston Globe*, 28 December 1929, pp. 1, 9; Orville Seagrave's position at WSAR is noted in *Variety Radio Directory*,

1940, p. 590; "Three Brothers Given Terms as Vagrants," *Boston Globe*, 27 July 1932, p. 7; "Pole Steals Air!," *Radio-Craft*, April 1940, p. 584.

21. Jill Hills, *The Struggle for Control of Global Communication: The Formative Century* (Urbana: University of Illinois Press, 2002), 200–206; Rebecca Scales, *Radio and the Politics of Sound in Interwar France, 1921–1939* (New York: Cambridge University Press, 2016), esp. 111–57; Smulyan, *Selling Radio*, 18; Douglas, *Listening In*, 80–82. "Third Annual Report of the Federal Radio Commission to the Congress of the United States Covering Period from October 1, 1928 to November 1, 1929," 78–79, https://docs.fcc.gov/public/attachments/DOC-338387A1.pdf.

22. Jennifer Leeman, "Categorizing Latinos in the History of the US Census: The Official Racialization of Spanish," José Del Valle, ed., *A Political History of Spanish: The Making of a Language* (New York: Cambridge University Press, 2013), 308–16.

23. Casillas, *Sounds of Belonging*, 40–41.

24. Sonia Robles, *Mexican Waves: Radio Broadcasting Along Mexico's Northern Border, 1930–1950* (Tucson: University of Arizona Press, 2019), 57, 63–64 (Baker quoted on p. 64); Gene Fowler and Bill Crawford, *Border Radio: Quacks, Yodelers, Pitchmen, Psychics, and Other Amazing Broadcasters of the American Airwaves*, rev. ed. (Austin: University of Texas Press, 2002), 67–102.

25. McChesney, *Telecommunications, Mass Media, and Democracy*, 97–100, 114, 141, 143, 148, 242–43, 260–65.

26. Johnson to the FCC, 16 March 1939, File 89–6, Box 365, FCC Records 1927–46. For more on Johnson, see Krysko, *American Radio in China*, 120.

27. Benedict Anderson, *Imagined Communities: Reflections on the Origins and Spread of Nationalism*, rev. ed. (New York: Verso Books, 1991), esp. 42–45. For the application of Anderson's ideas to broadcasting, see Douglas, *Listening In*, 23–24, and Hilmes, *Radio Voices*, 11–12.

28. Russo, *Points on the Dial*, esp. 2, 4, 19–26, 163–64; Casillas, *Sounds of Belonging*, esp. 8–10, 15, 24, 40–42; Kelman, *Station Identification*, esp. 12–13, 20, 28, 53, 81, 83.

29. McCarthy to FRC [*sic*], 21 September 1935, and Sullivan to FCC, 6 January 1939, Box 287, File 67–4, FCC Records 1927–46.

30. Baron, *The English-Only Question*, 1–3, 169; Wiley, "Continuity and Change in the Function of Language Ideologies in the United States," 73; Judt and Lacorne, "The Politics of Language," 9; Gerstle, *American Crucible*, 94.

31. Milliken to FRC, 23 November 1934, Box 287, File 67–4, FCC Records 1927–46.

32. Logan to FCC, 27 November 1934, Box 287, File 67–4, FCC Records 1927–46.

33. Jacobson, *Whiteness of a Different Color*, 98–99; Gerstle, *American Crucible*, 128–86, esp. 129, 136–38, 153, 185; Nicholas Sammond, *Babes in Tomorrowland: Walt Disney and the Making of the American Child, 1930–1960* (Durham, NC: Duke University Press, 2005), 46–47; Lary May, *Screening Out the Past: The Birth of Mass Culture and the Motion Picture Industry*, rev. ed. (Chicago: University of Chicago Press, 1983), esp. 3–4, 237–38.

34. Roger Daniels, *Coming to America: A History of Immigration and Ethnicity in American Life* (New York: Harper Collins, 1990), 296; Samuel Walker, *Presidents and Civil Liberties from Wilson to Obama: A Story of Poor Custodians* (New York: Cambridge University Press,

2014), 79–124; Jennifer Schuessler, "F.D.R.'s Library Takes a Hard Look at Race," *New York Times*, 3 August 2023, p. C1.

35. St. Clair to FRC, 6 April 1934, Box 287, File 67–4, FCC Records 1927–46.

36. Horten, *Radio Goes to War*, 66–86.

37. Gerstle, *American Crucible*, 156–85, esp. 157, 160–61; Zaragosa Vargas, *Proletarians of the North: A History of Mexican Industrial Workers in Detroit and the Midwest, 1917–1933* (Berkeley: University of California Press, 1993), 169–70.

38. "Our Treatment of Aliens," letter to the editor from William Bauer, *New York Times*, 25 August 1937, p. 20.

39. "New Foreign Policy: War Talks for 'Native Consumption'; Broadcasters Here Let Them Go By," *New York Times*, 19 November 1939, p. X12. Letters to the editor from Bell and Grove, *New York Times*, 16 June 1940, p. X10. "Radio Foreign Language Curbed," *New York Times*, 12 June 1940, p. 18.

40. "Foreign Language Programs," letter to the editor written by Groucho Marx, *Los Angeles Times*, 30 August 1940, p. A4. Groucho Marx, "What This Country Needs," *This Week Magazine* (published by the United Newspapers Magazine Corporation), 16 June 1947, p. 7; the version cited was included with that day's edition of the *Los Angeles Times*.

41. "Groucho, Chico, Harpo . . . And Karl?," *Washington Post*, 13 October 1998, p. A1.

42. "Foreign Language Programs," letter to the editor written by Groucho Marx, *Los Angeles Times*, 30 August 1940, p. II-4.

43. Groucho Marx, *Groucho and Me* (New York: B. Geis Associates, 1959), 29.

44. "Why Foreign Language Broadcasts?," letter to the editor by Capps, *Los Angeles Times*, 12 September 1940, p. II-4. Robert W. Capps is listed as a licensed amateur radio operator in a succession of directories published in the 1920s and 1930s; for example, see United States, Department of Commerce, Bureau of Navigation, *Amateur Radio Stations of the United States* (Washington, DC: US Government Printing Office, 1926), 152; *Radio Amateur Callbook*, Spring 1938, p. 114; *Radio Amateur Callbook*, Spring 1941, p. 113. Capps had evidently just relocated from San Diego to Los Angeles at the time he wrote his letter, with his and his station's registered location identified as the latter city by the early 1940s after earlier registrations had put him in San Diego.

45. For the National Council on Freedom from Censorship's petition, see "Radio Safeguard Asked: Recording of Foreign Language Broadcasts Urged on FCC," *New York Times*, 26 August 1940, p. 17. For the FCC's National Defense Plan, see "The FCC to Add 'Policemen' to Patrol Radio Under National Defense Program," *New York Times*, 7 July 1940, p. X8. For restrictions on granting licenses to noncitizens, see "New Rules Forestall Fifth Column Activities," *New York Times*, 16 June 1940, p. X10. For the Foreign Legion's condemnation of foreign language broadcasting, see p. X10. "Legion Stand Protested: Broadcasters Defend Foreign Language Programs," *New York Times*, 28 September 1941, p. 31. For consideration of the American Legion within its historical context, see "Fly Asserts Radio Will Remain Free: Tells Liberties Union that Nation Faces No Emergency that Warrants Federal Rule," *New York Times*, 13 February 1941, p. 22. In addition to its one million members, an additional half million plus belonged

to affiliated auxiliary groups. For membership figures and affiliated auxiliary groups, see "Legion Enrollment a Record," *New York Times*, 28 July 1940, p. 19. For the Legion's historical support of "100 percent Americanism," see Thomas A. Rumer, *The American Legion: An Official History* (New York: M. Evans, 1990), 115. See also Kelman, *Station Identification*, 20, 81–82; Gerstle, *American Crucible*, 88, 93.

46. Percentages pertaining to foreign-born residents in the United States and their citizenship status are from the 1920, 1930, and 1940 censuses, https://www2.census.gov/library/publications/decennial/1920/volume-3/41084484v3ch01.pdf, https://www2.census.gov/library/publications/decennial/1930/population-volume-2/16440598v2ch03.pdf, https://www2.census.gov/library/publications/decennial/1940/population-volume-2/33973538v2p1ch2.pdf.

47. Kelman, *Station Identification*, esp. 12–13.

48. Kristin Hoganson, "Stuff It: Domestic Consumption and the Americanization of the World Paradigm," *Diplomatic History* 30.4 (2006), 571–94, esp. 572. Warren I. Cohen, *The Asian American Century* (Cambridge, MA: Harvard University Press, 2001), 79–128. George Lipsitz, "Land of a Thousand Dances: Youth, Minorities, and the Rise of Rock and Roll," Lary May, ed., *Recasting America: Culture and Politics in the Age of Cold War* (Chicago: University of Chicago Press, 1989), 267–84.

49. For example, see "Anti-Immigrant Group Backs Helming's Bill," *The Citizen* (Auburn, New York), 15 January 2020, pp. A1, A5; "English Only Rhetoric," letter to the editor written by Juan Ochoa, *The Monitor* (McAllen, Texas), 13 September 2015, p. 110; "Breaking the Barrier: Prince George and Dinwiddie Counties Cope with English-Only Policy," *Progress Index* (Petersburg, Virginia), 29 December 2013, p. A1; "Lawsuit Revised to Stay in Tulsa: Official English Provision Voter ID Challengers Don't Want Oklahoma City Venue," *Daily Oklahoman* (Oklahoma City, Oklahoma), 10 December 2010, p. 4A; "Oklahoma's English-Only Success Prompts Delegates Hope for Maryland Law," *Star-Democrat* (Easton, Maryland), 16 November 2010, p. A5; "Gingrich Backs English Push as Official Language," *Washington Times*, 27 January 2007, p. A06; "'No Spanish on My Dime': English-Only Sentiment Grows Loud," *Atlanta Journal Constitution*, 11 December 2006, p. 1A; "Push for 'Official' English Heats up: Legislation Reacts to Immigration," *USA Today*, 9 October 2006, p. 1A; "Ban on Speaking Navajo Leads Cafe Staff to Sue," *New York Times*, 20 December 2002, p. A26; "Vote 'No' on Official English Initiative," letter to the editor written by Rebekah Martindale, *Times-Independent* (Moab, Utah), 12 October 2000, p. 7.

Chapter 2. "An Invasion by Radio Is Crossing the Mexican Border"

1. Coffman to Dill, 18 January 1932 and Averill to Dill, 18 January 1932, included with Dill to Under Secretary of State William Castle, 21 January 1932, 812.76-Brinkley/86, DSNA 1930–39. Many other complaints can be found in File 44–3, Box 192, FCC Records 1927–46. For backgrounds on Coffman and Averill, see "Harvester Employees Give Donations," *Hutchison News* (Hutchison, Kansas), 17 December 1921, p. 13; "First Concert Next Tuesday: Municipal Band to Play Outdoor Program Twice Weekly Dur-

ing Summer," *Hutchison News*, 13 May 1932, p. 16; "Nearly 200 Attend International Harvester Dance," *Hutchison News*, 12 December 1937, p. 7; "Harvey Averill Dies in Vinita; Funeral Today," *Daily Oklahoman* (Oklahoma City, Oklahoma), 18 November 1948, p. 7.

2. For example, see R. Alton Lee, *The Bizarre Careers of John R. Brinkley* (Lexington: University of Kentucky Press, 2002), esp. 153–81; Eric S. Juhnke, *Quacks and Crusaders: The Fabulous Careers of John Brinkley, Norman Baker, and Harry Hoxsey* (Lawrence: University Press of Kansas, 2002), 126–30; Fowler and Crawford, *Border Radio*, 55; Doerksen, *American Babel*, 90.

3. Martin Brückner, "Lessons in Geography: Maps, Spellers, and Other Grammars of Nationalism in the Early Republic," *American Quarterly* 51.2 (1999), 311–43; K. Jill Fleuriet, *Rhetoric and Reality on the U.S.-Mexican Border: Place, Politics, and Home* (New York: Palgrave Macmillan, 2021), esp. 2–3, 143, 275; Eidsheim, "Marian Anderson and 'Sonic Blackness' in American Opera," 644–45.

4. Transcript of Brinkley's broadcast over station XER, 2 December 1931, pp. 283–84, enclosed with correspondence from the Secretary of the Interior Lyman Wilbur to the Secretary of State Henry Stimson, 25 February 1932, 812.76-Brinkley/91, DSNA 1930–39.

5. For the ever-expanding range of afflictions that Brinkley claimed his operation could cure, see Juhnke, *Quacks and Crusaders*, 8.

6. For the references to acne, constipation, and gas, see the transcript of Brinkley's broadcast over station XER, 2 December 1931, pp. 264, 283, 287; for efforts to sell electric clocks, insurance policies, and candy, see the transcript of Brinkley's broadcast from 27 November 1931, pp. 235–37; transcripts are included with Wilbur to Stimson, 25 February 1932, 812.76-Brinkley/91, DSNA 1930–39. For the over-the-radio prescription business and religious items for sale, see Lee, *The Bizarre Careers of John R. Brinkley*, 75–78, 163.

7. Coffman to Dill, 18 January 1932 and Averill to Dill, 18 January 1932.

8. Dill to Castle, 21 January 1932, 812.76-Brinkley/86, DSNA 1930–39. For Dill's prominence in the efforts to remove Brinkley from the air, see Fowler and Crawford, *Border Radio*, 206.

9. Juhnke, *Quacks and Crusaders*, 1–10; Lee, *The Bizarre Careers of John R. Brinkley*, 1–62; Pope Brock, *Charlatan: America's Most Dangerous Huckster, the Man Who Pursued Him, and the Age of Flimflam* (New York: Crown Publishers, 2008), 6–81, 154, 162; Gerald Carson, *The Roguish World of Doctor Brinkley* (New York: Reinhart and Company, 1960), 11–49.

10. Fowler and Crawford, *Border Radio*, 27–39 ("Lines on the map" quote is from p. 28). Brinkley's negative assessment of Milford is from John R. Brinkley to Baird, Kurtz and Dobson (a Kansas City-based accounting firm), 14 September 1936, Folder 5, Box 1, John R. Brinkley Papers, Kansas State Historical Society, Topeka, Kansas (hereafter Brinkley Papers). See also "XER: The Sunshine Station Between the Nations" (promotional pamphlet), Folder 1, Box 4, Brinkley Papers. This and all subsequent dollar conversions for amounts from 1913 or later are done via the US Bureau of Labor Statistics CPI inflation calculator at https://www.bls.gov/data/inflation_calculator

.htm; for before 1913, https://www.in2013dollars.com/us/inflation. In this instance, in the context of Depression-era inflation, the calculations here reflect the conversion between January and November 1932.

11. "Establishment of a Radio Broadcasting Station in Nogales, Sonora," report submitted by Altaffer, 28 July 1930, 812.76/17, DSNA 1930–39.

12. US Ambassador J. Reuben Clark to Stimson, 16 June 1931, 812.76-Brinkley/10, DSNA 1930–39.

13. Clark to Stimson, 16 June 1931, 812.76-Brinkley/10, DSNA 1930–39. Clark to Castle, 28 August 1931, File 44–3, Box 192, FCC Records 1927–46.

14. Telegram from Brinkley to Curtis, 23 September 1931, enclosed with Curtis to the State Department's Chief of the Division of Mexican Affairs Herschel V. Johnson, 24 September 1931, 812.76-Brinkley/24; Memorandum of Conversation Between John E. Singleton Jr. (Milford, Kansas) and Mr. Tanis of the Division of Mexican Affairs, 6 October 1931, 812.76-Brinkley/32; Correspondence Conducted by the Department Regarding Attempted Broadcasting Mexico by Dr. John R. Brinkley, 7 October 1931, 812.76-Brinkley/33; Congressman James Strong to Walter Newton (Secretary to President Hoover), 10 October 1931, 812.76-Brinkley/42; DSNA 1930–39. On Curtis's connections to Brinkley in Kansas, including using the KFKB, see Lee, *The Bizarre Careers of John R. Brinkley*, 158. "Grudge station" remark is from Eugene Martin (Belmont, Kansas) to the Federal Radio Commission, 14 July 1931, File 44–3, Box 192, FCC Records 1927–46.

15. Among the many sources that cover this well-trodden ground is W. Dirk Raat and Michael M. Brescia, *Mexico and the United States: Ambivalent Vistas*, 4th ed. (Athens: University of Georgia Press, 2010).

16. "The March of Radio: News and Interpretation of Current Radio Events," *Radio Broadcast*, November 1928, p. 16. Jeffrey H. Smulyan, "Power to Some People: The FCC's Clear Channel Allocation Policy," *Southern California Law Review* 44.3 (1971), 811–47, esp. 816–17; James C. Foust, *Big Voices of the Air: The Battle over Clear Channel Radio* (Ames: Iowa State University Press, 2000), 31, 63; Fowler and Crawford, *Border Radio*, 147, 203, 212; Brock, *Charlatan*, 165.

17. Fowler and Crawford, *Border Radio*, 67–102, 134–58.

18. Boyle, "Radio Equipment" (Voluntary Report), 11 May 1923, 812.74/191, and Dudley G. Dwyer, "The Radio Situation in the Guadalajara Consular District," 12 August 1925, 812.74/201, DSNA 1910–29. See also Fowler and Crawford, *Border Radio*, 200–201; Brock, *Charlatan*, 165.

19. For receiver ownership, see Douglas, *Listening In*, 128, 131. On the Mexican market, see J. Justin Castro, *Radio in Revolution: Wireless Technology and State Power in Mexico, 1897–1938* (Lincoln: University of Nebraska Press, 2016); Fowler and Crawford, *Border Radio*, 201–3; James Schwoch, *The American Radio Industry and Its Latin American Activities, 1900–1939* (Urbana: University of Illinois Press, 1990), 106–7, 142.

20. Fowler and Crawford, *Border Radio*, 38, 203, 212; Brock, *Charlatan*, 165.

21. "Radio Station in the Yankee Tongue," 27 December 1931, *La Prensa*, copy of translated article sent by Clark to Stimson, 29 December 1931, 812.76-Brinkley/79;

"Mexican Radio Fans Rap 'Gland' Station," 28 December 1931, *Washington Post*, copy forwarded by the Department of State to the US Consul in Santillo, Mexico, 2 January 1932, 812.76-Brinkley/80; "Mexico Was Not Willing to Sign Radio Conventions," *El Universal*, 21 December 1932, copy of translated article enclosed with Clark to Stimson, 23 December 1932, 812.76-Brinkley/131; DSNA 1930–39. Information about the readership of *La Prensa* and *El Universal* is from Julio Moreno, *Yankee Don't Go Home! Mexican Nationalism, American Business Culture, and the Shaping of Modern Mexico, 1920–1950* (Chapel Hill: University of North Carolina Press, 2003), 61–62.

22. US Consul in Monterrey Edward I. Nathan, "Radio Broadcasting Stations Set Up in the Consular District of Monterrey," 17 March 1931, 812.76-Brinkley/7, and US Vice Consul at Piedras Negras Harold C. Wood, "New Radio Broadcasting Station Under Construction at Villa Acuna, Coahuila, Mexico," 20 August 1931, 812.76-Brinkley/13, DSNA 1930–39. For the fines levied against Brinkley, see Crawford and Fowler, *Border Radio*, 41.

23. Fowler and Crawford, *Border Radio*, 215; Carson, *Roguish World*, 200. On Cárdenas and expropriation of American land and oil holdings, see John F. Dwyer, "Diplomatic Weapons of the Weak: Mexican Policymaking during the US-Mexican Agrarian Dispute, 1934–1941," *Diplomatic History* 26.3 (2002), 375–95.

24. Castle to the US Embassy in Mexico, 18 October 1932, 812.76-Brinkley/113, and NBC vice president C. W. Horn to NBC president M. H. Aylesworth, 12 June 1931, 576. E 1/16, Box 2884, DSNA 1930–39.

25. Untitled memorandum from the Department of State's Treaty Division, 23 June 1933, 812.76-Brinkley/143; Chief of the Department of State's Treaty Division Charles M. Barnes to Assistant Secretary of State Wilbur J. Carr, 28 August 1933, 812.76-Brinkley/146; Federal Radio Commission Chair and head of US delegation in Mexico E. O. Sykes to the Secretary of State Cordell Hull, 2 August 1933, 576. E 1/328, Box 2885; DSNA 1930–39. See also Fowler and Crawford, *Border Radio*, 211–13; Lee, *The Bizarre Careers of John R. Brinkley*, 164–65.

26. Fowler and Crawford, *Border Radio*, 55; Doerksen, *American Babel*, 90. For one such supportive letter, see R. Newby to Brinkley, 16 December 1932, Folder 3, Box 1, Brinkley Papers. "Medicine: XER Silenced," *Time*, 5 March 1934, http://content.time.com/time/subscriber/article/0,33009,747109,00.html.

27. Department of State to Clark, 21 November 1931, 812.76-Brinkley/67, DSNA 1930–39. Ultimately, the FCC and the State Department preserved over a thousand pages of documentation devoted to complaints on interference generated by Brinkley. Douglas, *Listening In*, 23–24, 133, 161–62.

28. Keagy to FRC, 1 November 1931 and Lester P. Huey (St. Louis, Missouri) to the Federal Radio Commission, 2 November 1931, forwarded by the Federal Radio Commission to the Department of State, 812.76-Brinkley/61; Mrs. Odo J. Sniff (Hollywood, California) to FRC, 27 October 1931 and Mrs. E. C. Feldhaus (Sydney, Montana) to FRC, undated, forwarded by the FRC to the Department of State, 812.76-Brinkley/63; the Lions Club of Newcastle, Wyoming, to the Federal Radio Commission [*sic*], 12 February 1936, forwarded by the FRC to the Department of State, 812.76-Brinkley/222; DSNA 1930–39.

29. Francisco E. Balderrama and Raymond Rodriguez, *Decade of Betrayal: Mexican Repatriation in the 1930s*, rev. ed. (Albuquerque: University of New Mexico Press, 2006), esp. 1–3 ("frenzy" direct quote is from p. 2). Swindler to WSB, 10 November 1931, Jenkins to WSB, 27 November 1931 (both letters forwarded by WSB to FRC, in turn forwarded to the Department of State), 812.76-Brinkley/88, and Taylor to WSB, 17 October 1931 (forwarded by WSB to the Federal Radio Commission, in turn forwarded to the Department of State), 812.76-Brinkley/64, DSNA 1930–39. KMMJ owner H. H. Johnson to FRC, 14 October 1931, File 44–3, Box 192, FCC Records, 1927–46.

30. "Fight Radio Station on Mexican Border," *New York Times*, 28 December 1931, p. 13; "Strumming Those Symptoms—the Doc's (Mex) on the Air," *Chicago Daily News*, 12 March 1937, p. 10, clipping of article in Box 1, Folder 7, Arthur Carson Papers, Kansas State Historical Society, Topeka, Kansas; "Radio Giant," *Time*, 14 May 1934, p. 46; Mrs. E. C. Feldhaus to the FRC, undated, forwarded by the FCC to the Department of State, 812.76-Brinkley/63, DSNA 1930–39 (emphasis added).

31. Douglas, *Listening In*, 27, 100–123 (quote from p. 27); Guntram Henrick Herb, *Under the Map of Germany: Nationalism and Propaganda, 1918–1945* (New York: Routledge, 1997), 7–8, 178; Norman J. W. Thrower, *Maps and Civilization: Cartography in Culture and Society*, 2nd ed. (Chicago: University of Chicago Press, 2008), 1, 217; Carl F. Kaestle, *Pillars of the Republic: Common Schools and American Society, 1780–1860* (New York: Harper Collins, 1983), 6–7, 99–101; Wayne J. Urban and Jennings L. Wagoner Jr., *American Education: A History*, 4th ed. (New York: Routledge, 2009), 62–63, 92–93; William J. Reese, *History, Education, and the Schools* (New York: Palgrave-MacMillan, 2007), 82–89, 145–50.

32. Michael Heffernan, "The Cartography of the Fourth Estate: Mapping the New Imperialism in British and French Newspapers, 1875–1925," James R. Ackerman, ed., *The Imperial Map: Cartography and the Mastery of Empire* (Chicago: University of Chicago Press, 2009), 265. See also Russo, *Points on the Dial*, 17–18.

33. Russo, *Points on the Dial*, 17–26.

34. Mrs. R. J. Killion (Lyons, Kansas) to KMMJ, 7 October 1941, forward by KMMJ owner to FRC Secretary James Baldwin, 19 October 1931; Trapp to Baldwin, 6 September 1932; correspondence from File 44–3, Box 192, FCC Records, 1927–46; Conway to WSB (forwarded to the FRC, in turn forwarded to the Department of State), 812.76-Brinkley/64, DSNA 1930–39. Examples of listeners castigating Brinkley's broadcasts as a "menace" forwarded to the Department of State include R. H. Jackson Jr. (Madison, Wisconsin) to FRC, 7 December 1931, 812.76-Brinkley/73, and H. C. R. Stewart (Columbia, Missouri) to the FRC, 15 December 1931, 812.76-Brinkley/88. See also Barnes to Carr, 28 August 1933, 812.76-Brinkley/146, DSNA 1930–39.

35. The FRC forwarded the following letters to the Department of State and they are filed in DSNA 1930–39: Stewart to the FRC, undated, 812.76-Brinkley/88; Williams to the FRC, 22 January 1932, 812.76-Brinkley/88; Milburn to FRC, 22 January 1932, 812.76-Brinkley/88; U. S. A. Radio Fan to FRC, 16 November 1931, 812.76-Brinkley/69. For Stewart's employment, see Missouri, *Appendix to the House and Senate Journals of the General Assembly of State of Missouri* (Jefferson City, MO: Midland Printing Company, 1933), 8.

36. Population statistics are from the 1930 United States Census, https://www2.census.gov/library/publications/decennial/1930/population-volume-1/03815512v1.zip.

37. Fowler and Crawford, *Border Radio*, 43; Casillas, *Sounds of Belonging*, 40.

38. "Appendices and Tables: Allocation Provisions of the Havana Agreement," *Broadcasting Yearbook* (1941), 403; Foust, *Big Voices of the Air*, 64–65; Fowler and Crawford, *Border Radio*, 222–23; Lee, *The Bizarre Careers of John R. Brinkley*, 177–78.

39. Fowler and Crawford, *Border Radio*, 223; Lee, *The Bizarre Careers of John R. Brinkley*, 178; Moreno, *Yankee Don't Go Home*, 53–54; Dwyer, "Diplomatic Weapons of the Weak," 375–85. The date of ratification by the US Senate is noted in United States, *Limit Power of Radio Stations: Hearings Before the Committee on Interstate and Foreign Commerce United States Senate, Eightieth Congress, Second Session on S. 2231* (Washington, DC: US Government Printing Office, 1948), 96. F. Lee Armstrong to President Franklin D. Roosevelt, 10 January 1938, 812.76-Brinkley/264, DSNA 1930–39.

40. Martin L. Dugow (Manhattan, New York) to FRC [*sic*], 20 February 1939; Birkhead to FCC Chairperson James W. Fly, 29 August 1940; E. H. Hester (Arcadia, Louisiana) to FRC [*sic*], 29 January 1938; correspondence from Box 192, File 44–3, FCC Records 1927–46. For biographical information on Birkhead, see "Dictionary of Unitarian and Universalist Biography," https://www.uudb.org/birkhead-leon-milton.

41. Fowler and Crawford, *Border Radio*, 224–25; Lee, *The Bizarre Careers of John R. Brinkley*, 178; Moreno, *Yankee Don't Go Home*, 53–58. For the broader context of US-Mexican relations, the tensions that enveloped it, and Mexico's effectiveness at securing concessions despite its relative weakness vis-à-vis the United States, see Dwyer, "Diplomatic Weapons of the Weak," esp. 386–95.

42. John Brinkley to Minerva Brinkley, 21 July 1941, Folder 1, Box 1, and John Brinkley to Attorney Wallace Davis, 10 January 1942, Folder 8, Box 1, Brinkley Papers. Lee, *The Bizarre Careers of John R. Brinkley*, 219–24.

43. Doerksen, *American Babel*, 90; Fowler and Crawford, *Border Radio*, 55. For farmer opposition to the loss of their favored stations as American radio commercialized and nationalized, see Razlogova, *The Listener's Voice*, 34.

Chapter 3. "To Help the French Speaking People of Louisiana"

1. Griffith to Rockefeller Foundation General Education Board Member W. W. Brierley, 6 June 1938 and Griffith to Rockefeller Foundation General Education Board field agent Leo M. Favrot, 6 June 1938, Folder 3708, Box 360, Series 1.2 (Appropriations Secondary and Higher Education), General Education Board Appropriations Records, Rockefeller Archive Center, Sleepy Hollow, New York (hereafter General Education Board Appropriations Records).

2. Smith to Griffith, 14 January 1939, and Smith to the Rockefeller Foundation's Director of Humanities Programs John Marshall, 17 January 1939, Folder 3708, Box 360, General Education Board Appropriations Records (a copy of Smith's 14 January 1939 letter can also be found in Folder 24, Box 33, Office of the Chancellor Records,

Special Collections, Hill Memorial Library, Louisiana State University, Baton Rouge, Louisiana (hereafter Office of the Chancellor Records at LSU)).

3. Harley Albert Smith, "A Recording of English Sounds at Three Age Levels in Ville Platte, Louisiana" (PhD dissertation: Louisiana State University, 1936). Louise Olivier, "A Glossary of Variants from Standard-French in St. Landry Parish" (MA thesis, Louisiana State University, 1937).

4. "General Plan for the Organization and Operation of Radio and Public Address Department at Louisiana State University," undated (date stamp indicating 31 May 1938 receipt at Rockefeller Foundation), Folder 3708, Box 360, General Education Board Appropriations Records.

5. Smith to the Rockefeller Foundation's Director of Humanities Programs John Marshall, 17 January 1939, Folder 3708, Box 360, General Education Board Appropriations Records. Quotes are from Clara Krefting Mawhinney and Harley A. Smith, *Business and Professional Speech* (New York: American Book Company, 1950), 5–6, and Harley A. Smith and Ida Lee King Wilbert, *Everyday English* (Austin, TX: Steck-Vaughn Company, 1965), 2. The Steck-Vaughn Company published a second edition of *Everyday English* in 1974, which contained the same quote on the same page as the 1965 edition.

6. Smith to Marshall, 17 January 1939; Smith to Griffith, 14 January 1939, Folder 3708, Box 360, General Education Board Appropriations Records.

7. Hugh R. Slotten, *Radio's Hidden Voice: The Origins of Public Broadcasting in the United States* (Urbana: University of Illinois Press, 2009), 9–79; McKinley Mayes, "Status of Agricultural Research Programs at 1890 Land-Grant Institutions and Tuskegee University," 53, and Dan Godfrey and Alton Franklin, "Extension Programs at the 1890 Land-Grant Institutions," 61, 64, in Ralph D. Christy and Lionel Williamson, eds., *A Century of Service: Land-Grant Colleges and Universities, 1890–1990* (New Brunswick, NJ: Transaction Publishers, 2012).

8. Josh Shepperd, "Infrastructure of the Air," esp. 231, 234–35.

9. Shepperd, "Infrastructure of the Air," 235–37; McChesney, *Telecommunications, Mass Media, and Democracy*, 47–48, 232; Slotten, *Radio's Hidden Voice*, esp. 160–66; Paul Saettler, *The Evolution of American Education Technology* (Greenwich, CT: Information Age Publishing, 2004), 215.

10. Rockefeller Foundation 1937 Annual Report, accessed at https://www.rockefeller foundation.org/wp-content/uploads/Annual-Report-1937-1.pdf; Raymond Fosdick, *The Story of the Rockefeller Foundation* (New York: Harper and Row, 1952). Hilmes, *Network Nations*, 107–15; Douglas, *Listening In*, 130, 141; Josh Shepperd, *Shadow of the New Deal: The Victory of Public Broadcasting* (Urbana: University of Illinois Press, 2023), 60–73; Shepperd, "Infrastructure in the Air"; Josh Shepperd, "Rockefeller Underwriting of Local, Regional, and National Educational Broadcasting Experiments, 1934–1940," unpublished paper (2013), https://www.issuelab.org/resources/28023/28023.pdf. See also Steven C. Wheatley, "Introduction," in Raymond Fosdick, *The Story of the Rockefeller Foundation* (New York: Routledge, 2017), vii-xvii.

11. "General Plan for the Organization and Operation of Radio and Public Address Department at Louisiana State University," p. 3, undated (date stamp indicating 31 May 1938 receipt at Rockefeller Foundation), Folder 3708, Box 360, General Education Board Appropriations Records.

12. Untitled document (contains paragraph summation of project objectives), 24 March 1938; memorandum of conversation between Rockefeller Foundation officials and LSU Director of Extension P. H. Griffith, 11 April 1938; Griffith to Louisiana State University president James Monroe Smith, 28 April 1938; General Education Board, "Grant In Aid—Southern Program—Public Education" (documents the decision to fund the project), 18 May 1938; "General Plan for the Organization and Operation of Radio and Public Address Department at Louisiana State University," undated (received 31 May 1938); Folder 3708, Box 360, General Education Board Appropriations Records.

13. "The Problem of Educational Broadcasting and a Plan for Its Solution," 5 March 1935, Folder 3394, Box 284, Series 200R, Record Group 1.1, Rockefeller Foundation Records (Projects), Rockefeller Archive Center, Sleepy Hollow, New York (hereafter Rockefeller Foundation Project Records). Smith to Marshall, 17 January 1939; "The Limitations of the Radio as a Means of Classroom Instruction," enclosed with NACRE Director Levering Tyson to Cline N. Koon of the US Office of Education, 7 November 1932, Folder "Office of Education," Box 3, Records of the National Advisory Council on Radio in Education, New York Public Library, New York, New York. Smith to Griffith, 14 January 1939, Folder 3708, Box 360, General Education Board Appropriations Records. For the NCER and NACRE rivalry, see Slotten, *Radio's Hidden Voice*, 164–65; Eugene Leach, "Tuning out Education: The Cooperation Doctrine in Radio, 1922–38," *Current*, August 1983, esp. 3, 7, reprint accessed online via Education Resources Information Center (ERIC) database, https://files.eric.ed.gov/fulltext/ED248835.pdf; "Eighth Institute for Education by Radio," *Education By Radio: A Bulletin to Promote the Use of Radio for Educational, Cultural, and Civic Purposes* 7.6 (1937), 25; "The Radio Panorama," *Education By Radio: A Bulletin to Promote the Use of Radio for Educational, Cultural, and Civic Purposes* 7.7 (1937). NCER published *Education By Radio*.

14. Carl L. Bankston and Jacques Henry, "Endogamy among Louisiana Cajuns: A Social Class Explanation," *Social Forces* 77.4 (1999), 1321; Mark F. Dewitt, "From Chanky-Chank to Yankee Chanks: The Cajun Accordion as Identity Symbol," Helena Simonett, ed., *The Accordion in the Americas: Klezmer, Polka, Tango, Zydeco, and More!* (Urbana: University of Illinois Press, 2012), 45; Thomas A. Klingler, "How Much Acadian Is There in Cajun?," Ursula Mathis-Moser and Günther Bischof, eds., *Acadians and Cajuns: The Politics and Culture of French Minorities in North America* (Innsbruck, Austria: University of Innsbruck Press, 2009), 94–95; Sylvie Dubois and Megan Melançon, "Cajun Is Dead—Long Live Cajun: Shifting from a Linguistic to a Cultural Community," *Journal of Sociolinguistics* 1.1 (1997), 65–66. See also Noah Arceneaux, "Acadian Airwaves: A History of Cajun Radio," *Journal of Radio and Audio Media* 30.2 (2023).

15. Ryan André Brasseaux, *Cajun Breakdown: The Emergence of an American-Made Music* (New York: Oxford University Press, 2009), 123; Barry Jean Ancelet, Jay D. Edwards,

and Glen Pitre, *Cajun Country* (Jackson: University of Mississippi Press, 1991), 49–50. For group listening in other contexts, see Cohen, *Making a New Deal*, 129–42, esp. 133–34; Krysko, *American Radio in China*, 80–83; Scales, *Radio and the Politics of Sound in Interwar France, 1921–1939*, 29–34, 72; Castro, *Radio in Revolution*, 174–76; Fowler and Crawford, *Border Radio*, 201–2; Alejandra Bronfman, *Isles of Noise: Sonic Media in the Caribbean* (Chapel Hill: University of North Carolina Press, 2016), 101, 112–13.

16. Memorandum of conversation between Marshall and Griffith, 1 March 1938; Griffith to Favrot, 8 March 1938; and "General Plan for the Organization and Operation of Radio and Public Address Department at Louisiana State University," undated (received 31 May 1938); Folder 3708, Box 360, General Education Board Appropriations Records. Direct quote is from the "General Plan."

17. Hilmes, *Network Nations*, 110–11; Slotten, *Radio's Hidden Voice*, 194; Rockefeller Foundation biography of John Marshall, https://rockfound.rockarch.org/biographical/-/asset_publisher/6ygcKECNI1nb/content/john-marshall; William J. Buxton, "John Marshall and the Humanities in Europe: Shifting Patterns of Rockefeller Foundation Support," *Minerva* 41.2 (2003), esp. 135, 142–44; Shepperd, *Shadow of the New Deal*, 66–69; Shepperd, "Rockefeller Underwriting of Local, Regional, and National Educational Broadcasting Experiments, 1934–1940," 5. Jefferson Pooley, "The New History of Mass Communication Research," David W. Park and Jefferson Pooley, eds., *The History of Media and Communication Research: Contested Memories* (New York: Peter Lang, 2008), 49–51; Seth Finn, "Office of Radio Research," Christopher Sterling and Cary O'Dell, eds., *The Concise Encyclopedia of American Radio* (New York: Routledge, 2010), 534.

18. General Education Board, "Grant In Aid—Southern Program—Public Education" (documents the decision to fund the project), 18 May 1938, Folder 3708, Box 360, General Education Board Appropriations Records.

19. Transcript of Emeritte O. Perret (cousin of Louise Olivier) Oral History Interview, interviewed by Pamela Rabalais and Yvonne Olivier, 12 August 1995, pp. 3, 24, 34–35 of transcript (includes all direct quotes except "life of any party"), Tape 735, 4700.0513, Acadian Handicraft Project, Special Collections, Hill Memorial Library, Louisiana State University, Baton Rouge, Louisiana (hereafter Acadian Handicraft Project Records). Mary Alice Fontenot, "En Passant," *Daily World* (Opelousas, Louisiana), 16 July 1965, p. 6 (includes "life of any party" quote). Birthplace and education information from W. Fitzhugh Brundage, "Le Reveil de la Louisiane," Brundage, ed., *Where These Memories Grow: History, Memory, and Southern Identity* (Chapel Hill: University of North Carolina Press, 2000), 286–87.

20. Jacques Henry, "The Louisiana French Movement: Actors and Actions in Social Change," 187, and Becky Brown, "The Development of a Louisiana French Norm," 222, Arthur Valdman, ed., *French and Creole in Louisiana* (New York: Plenum Press, 1997); W. Fitzhugh Brundage, "Memory and Acadian Identity, 1920–1960: Susan Evangeline Walker Anding, Dudley LeBlanc, and Louise Olivier, or the Pursuit of Authenticity," Mathis-Moser and Bischof, eds., *Acadians and Cajuns*, 65–67.

21. Transcript of O. Perret Oral History Interview, p. 33 of written transcript, Acadian Handicraft Project Records.

22. Dubois and Melançon, "Cajun Is Dead—Long Live Cajun," 66; Dewitt, "From Chanky-Chank to Yankee Chanks," 45. Brundage, *Where These Memories Grow*, 287; Alex Melancon, "A Tip on Calling Them Cajuns," *Clarion-News* (Opelousas, Louisiana), 29 October 1942, p. 1; Brasseaux, *Cajun Breakdown*, 127.

23. Dubois and Melançon, "Cajun Is Dead—Long Live Cajun," esp. 77; Virginia R. Dominguez, *White By Definition: Social Classification in Creole Louisiana* (New Brunswick, NJ: Rutgers University Press, 1986), 12–15, 149–81. See also Arceneaux, "Acadian Airwaves."

24. Klingler, "How Much Acadian Is There in Cajun?," esp. 97.

25. George Thomas, *Linguistic Purism* (London: Longman, 1991), 101, 112; John Edwards, *Language and Identity: An Introduction* (New York: Cambridge University Press, 2009), 5; Brundage, "Le Reveil de la Louisiane," 287.

26. Smith to General Extension Division Director P. H. Griffith, 14 January 1939, and Smith to John Marshall, 17 January 1939, Folder 3708, Box 360, General Education Board Appropriations Records; Donald Winford, "Ideologies of Language and Socially Realistic Linguistics," Sinfree Makoni, Geneva Smitherman, Arnetha F. Ball, and Arthur K. Spears, eds., *Black Linguistics: Language, Society, and Politics in Africa and the Americas* (New York: Routledge Press, 2003), 31–32.

27. Louise Olivier, "French Radio Project: Report of the Activities of the Field Worker," undated, received 26 May 1939, Folder 3708, Box 360, General Education Board Appropriations Records.

28. Edwards, *Linguistic Purism*, 5. See also Brian Weinstein, "Francophonie: Purism at the International Level," Björn H. Jernudd and Michael J. Shapiro, eds., *The Politics of Language Purism* (New York: Mouton de Gruyter, 1989), 59; Robert McColl Millar, *Language, Nation and Power* (New York: Palgrave MacMillan, 2005), 19; Thomas, *Linguistic Purism*, 1–2, 49, 101, 112; Winford, "Ideologies of Language and Socially Realistic Linguistics," 23, 27–28, 32, 164.

29. Memorandum of conversation between Marshall and Griffith, 2 March 1938 (addresses President Smith's role in reorganizing LSU's radio initiatives under the purview of the Extension Division and Griffith); Griffith to General Education Board field agent Leo M. Favrot, 8 March 1938; Griffith to President Smith, 28 April 1938; Memorandum by Favrot, 6 May 1938 (indicates that the request for funding had been submitted by President Smith); Rockefeller Foundation General Education Board Secretary W. W. Brierley to President Smith, 26 May 1938; Griffith to Brierley, 6 June 1938 (indicates that it was President Smith who hired Olivier); Folder 3708 (Louisiana State University-Radio, 1938–1941), Box 360, General Education Appropriations Board Records. President Smith to Favrot, 3 May 1938; President Smith to Favrot, 10 May 1938; President Smith to Brierly, 9 June 1938; President Smith to President of the Baton Rouge Broadcasting Company Charles P. Manship Jr., 12 August 1938; Folder 24 ("General Education Board—French Radio Project 1938–1940"), Box 33, Office of the Chancellor Records at LSU. For President Smith's outreach to various Louisiana radio stations seeking cooperation in broadcasting the university's educational programs, see Folder 12, Box 38, Office of the Chancellor Records at LSU.

30. Smith to Griffith, 14 January 1939 (from where the direct quote is taken) and memorandum of interviews with Harley Smith and LSU Director of Radio Programs Ralph Steetle, 31 January 1939, Folder 3708, Box 360, General Education Board Appropriations Records.

31. Griffith to President Smith, 10 February 1939 (quotes Griffith's favorable assessment of Olivier consulting "prominent citizens"); H. Smith to Griffith, 14 January 1939; H. Smith to Marshall, 17 January 1939; memorandum of interviews with H. Smith and Steetle, 31 January 1939; excerpts from listeners, enclosed with Smith's letters to Griffith (14 January 1939) and Marshall (17 January 1939); Folder 3708, Box 360, General Education Board Appropriations Records. Harley Smith's resignation letter to Griffith dated 14 January 1939 and Griffith's letter to President Smith dated 10 February 1939 also filed in Folder 24, Box 33, Office of the Chancellor Records at LSU.

32. Memorandum interviews with Smith and Steetle, 31 January 1939 (notes the shift to standard French after five broadcasts), and Griffith to President Smith, 10 February 1939 (for "correct French" quote), Folder 3708, Box 360, General Education Board Appropriations Records. See also Smith to President Smith, 20 January 1936, Folder 24, Box 33, Office of the Chancellor Records at LSU (contains H. Smith's quote refusing to accept responsibility for the change to using standard French in the broadcasts).

33. Favrot, Interoffice Correspondence, "Dr. Harley Smith's Letter January 17, 1939," 23 January 1939, Folder 3708, Box 360, General Education Board Appropriations Records.

34. Memorandum of Interview with Dr. Harley Smith and Dr. Ralph Steetle, "On French-speaking program at Louisiana State University," 31 January 1939, Folder 3708, Box 360, General Education Board Appropriations Records.

35. Griffith to President Smith (for quote pertaining to Steetle's confidence), 10 February 1939, and Griffith to Favrot (for quote praising Olivier), 20 May 1939, Folder 3708, Box 360, General Education Board Appropriations Records. Griffith's 10 February 1939 letter also filed in Folder 24, Box 33, Office of the Chancellor Records at LSU.

36. "French Radio Project: Report of the Activities of the Field Worker," undated, received 26 May 1939, Folder 3708, Box 360, General Education Board Appropriations Records.

37. "French Radio Project: Report of the Activities of the Field Worker," undated, received at the Rockefeller Foundation on 26 May 1939, Folder 3708, Box 360, General Education Board Appropriations Records. "French Renaissance," *The Assumption Pioneer* (Napoleonville, Louisiana), 8 April 1939, p. 2.

38. Sammy Toyoki and Andrew Brown, "Stigma, Identity, and Power: Managing Stigmatized Identities through Discourse," *Human Relations* 67.6 (2014), 715–37, esp. 721; Elena Doldor and Doyin Atewologun, "Why Work It When You Can Dodge It? Identity Responses to Ethnic Stigma Among Professionals," *Human Relations* 74.6 (2021), esp. 892, 898; Linda Morrice, "Refugees in Higher Education: Boundaries of Belonging and Recognition, Stigma and Exclusion," *International Journal of Lifelong Education* 23.5 (2015), esp. 653, 666; Holly S. Slay and Delmonize A. Smith, "Professional

Identity Construction: Using Narrative to Understand the Negotiation of Professional and Stigmatized Cultural Identities," *Human Relations* 64.1 (2011), 85–107, esp. 86, 101.

39. Brundage, "Memory and Acadian Identity," 66–68; Casillas, *Sounds of Belonging*, esp. 38; Bonfiglio, *Race and the Rise of the Standard American*, 115, 236; Leeman, "Categorizing Latinos in the History of the US Census," 313–16; Stephanie Lindemann, "Who Speaks 'Broken English'? US Undergraduate Perceptions of Non-Native English," *International Journal of Applied Linguistics* 15.2 (2005), 204–5.

40. "French Radio Project: Report of the Activities of the Field Worker," undated, received 26 May 1939 (contains Mann's quote), and Favrot to Griffith, 6 June 1939 (contains request for revised report), Folder 3708, Box 360, General Education Board Appropriations Records.

41. "Jimmy the Stooge," *Time*, 10 July 1939, pp. 16–17. "To Continue Probe on Use of LA Property," *Daily Advertiser* (Lafayette, Louisiana), 14 June 1939, p. 1; "Smith Plans to Aid in Probe WPA Charges: L. S. U. President Promises WPA Leader Cooperation in Planned Investigation," *Daily Advertiser* (Lafayette, Louisiana), 20 June 1939, pp. 1, 2; "Police Find No Trace of Dr. Smith," *Alexandria Daily Town Talk* (Alexandria, Virginia), 26 June 1939, p. 1; "Reward Offered for Smith's Arrest," *Alexandria Daily Town Talk* (Alexandria, Virginia), 27 June 1939, p. 1; "Louisiana U. Head Gives Up: Officer Flying to Canada to Get Dr. Smith," *Chicago Sunday Tribune*, 2 July 1939, p. 1; "Missing Louisiana U. Head Arrested in Canada, Will Face Embezzlement Charge: Smith and Wife Held in Jail at Brockville, Ont.," *St. Louis Post Dispatch*, 2 July 1939, pp. 1, 2. See also finding aid to James Monroe Smith Papers, p. 4, Special Collections, Hill Memorial Library, Louisiana State University (also available at https://www.lib.lsu.edu/sites/default/files/sc/findaid/4490.pdf); Michael G. Wade, "Villainy, Virtue, and Louisiana Political Culture: Paul Herbert and the Augean Stables at LSU, 1939–1941," *Louisiana History: The Journal of the Louisiana Historical Association* 49.1 (2008).

42. Griffith to Favrot, 14 June 1939 (including the attachment from Steetle entitled "Louisiana French Program" and a radio script entitled "Handling, Packing, and Marketing of Spring Vegetables"), Folder 3708, Box 360, General Education Board Appropriations Records.

43. Griffith to Favrot, 14 June 1939.

44. "Condensed Statement on the General Extension Division," enclosed with P. H. Griffith to Paul Herbert, Acting President of Louisiana State University, 20 December 1939, Folder 9, Box 6, Office of the Chancellor Records at LSU.

45. Claire L. Gueymard, "Songs Acadians Sang: Old French Melodies Are Heard Again Among the Bayous of Louisiana," *New York Times*, 29 March 1942, p. XX6, copy of article on file in Folder 3708, Box 360, General Education Board Appropriations Records.

46. Brundage, "Memory and Acadian Identity," 64–66 (both Olivier's and Brundage's quotes are on p. 64).

47. Quoted in Brundage, "Memory and Acadian Identity," 66.

48. Brundage, "Memory and Acadian Identity," 66. "Miss Louise Olivier, Promoter of Native Acadian Folk Lore, Dies in Arnaudville," *Teche News* (St. Martinville, Louisiana), 26 July 1962, p. 1.

49. Weinstein, "Francophonie: Purism at the International Level," 59 (contains "red neck" quote; The author cited identified CODOFIL incorrectly as the "Committee for the Defense of French in Louisiana." Thanks to Noah Arceneaux for flagging the error.); Olivia Walsh, *Linguistic Purism: Language Attitudes in France and Quebec* (Philadelphia: John Benjamins Publishing Company, 2016); Herman Lebovics, *True France: The Wars over Cultural Identity, 1900–1945* (Ithaca, NY: Cornell University Press, 1992), esp. 98–134; Richard Kuisel, *Seducing the French: The Dilemma of Americanization* (Berkeley: University of California Press, 1993), esp. 41, 192–93, 220; Richard Kuisel, *The French Way: How France Embraced and Rejected American Values and Power* (Princeton, NJ: Princeton University Press, 2012), esp. 60, 71–72, 162, 265, 308–13, 330; Nils Langer, *Linguistic Purism in Action: How Auxiliary Tun Was Stigmatized in Early New High German* (New York: Walter de Gruyter, 2001); Khawla Badwan, *Language in a Globalised World: Social Justice Perspectives on Mobility and Contact* (New York: Palgrave Macmillan, 2021); Millar, *Language, Nation, and Power*.

50. The editorial board, "Ebonics Proposal Not Good Solution"; Ellen Goodman, "Ebonics a Form of Defeatism"; Leonard Larson, "Ebonics Decision Harmful to Blacks"; and Clarence Page, "Speaking Properly Has No Substitute"; all editorials from *Alexandria Daily Town Talk* (Alexandria, Louisiana), 29 December 1996, p. 22. Barbara Birch, "Ebonics: The Debate Which Never Happened," *Ethnic Studies Review* 22.1 (1999), 47. See also John Baugh, *Beyond Ebonics: Linguistic Pride and Racial Prejudice* (New York: Oxford University Press, 2000), and Theresa Perry and Lisa Delpit, eds., *The Real Ebonics Debate: Power, Language, and the Education of African-American Children* (Boston: Beacon Press, 1998).

51. Harley Smith's 1936 dissertation, for example, was cited as recently as 2018 in Sylvie Dubois, Emilie Gagnet Leumas, and Malcolm Richardson, *Speaking French in Louisiana, 1720–1955: Linguistic Practices of the Catholic Church* (Baton Rouge: Louisiana State University Press, 2018).

Chapter 4. "An Efficient Way to Spread Shakespeare's Beautiful Language"

1. "In the Classroom and on the Campus," *New York Times*, 6 December 1936, p. N5. See also "Radio Programs-Development-Boston," RF 35118, File 3513, Box 295, Rockefeller Foundation Project Records.

2. "Plans World Radio as Aid to Good Will," *New York Times*, 15 June 1931, p. 24 (includes the direct quote); T. R. Kennedy Jr., "Friendship 'Bridge': WRUL, Idea Born at Versailles in 1919, Urges World Amity on Powerful Waves," *New York Times*, 19 January 1941, p. X12; Jerome Berg, *On the Shortwaves, 1923–1945: Broadcast Listening in the Pioneer Days of Radio* (Jefferson, NC: McFarland and Company, 1999), 58–59. "New Airwave Used to Bind Americas," *New York Times*, 16 February 1938, p. 11.

3. "Foundation Broadcasts Messages of Hope to Oppressed People Abroad," *New York World-Telegram*, 30 January 1941 (contains "common language tie" quote), copy of article in Folder 3533, Box 296, Rockefeller Foundation Project Records. Lemmon to Arthur Jones (Council of National Defense, State Department), 2 October 1940 (contains "very permanent value" quote), Folder 1189; Lemmon to Laurance

Rockefeller, 14 August 1940, Folder 1189, and "Progress Report of 1950 Operations" (contains "opportunities for closer inter-American understanding" quote), Folder 1192, Box 135, Office of the Messrs. Rockefeller Records, Cultural Interests, Series E, Radio-World Wide Broadcasting Foundation, Rockefeller Archive Center, Sleepy Hollow, New York (hereafter Messrs. Rockefeller Cultural Interests Records). "Plans World Radio as Aid to Good Will," *New York Times*, 15 June 1931, p. 24.

4. Krysko, *American Radio in China*, 90–125, esp. 96.

5. Donald H. Short, "Woodrow Wilson First 'DX' Broadcaster: New York Radio Man Recalls Incidents of 1919," *Times Leader Evening News* (Wilkes Barre, Pennsylvania), 16 February 1924, p. 4 (includes quote about Wilson's views on radio's future); "Walter Lemmon, Inventor, Is Dead," *New York Times*, 21 March 1967, p. 46. See also "Story of W1XAL," 6 May 1935, Folder 3513, Box 295, and Thomas Kellend, "Foundation Broadcasts Messages of Hope to Oppressed Peoples Abroad," *New York World-Telegraph*, 30 January 1941, Folder 3533, Box 296, Rockefeller Foundation Project Records.

6. Eric Underwood, "The Radio University in Peace and War," *American Scholar* 14.1 (1944), 88–89.

7. Headrick, *The Invisible Weapon*, 202–3; Daniel R. Headrick, "Shortwave Radio and Its Impact on International Telecommunications Between the Wars," *History and Technology* 11.1 (1994), 21–32, esp. 23–26.

8. *EMF Electrical Year Book: An Encyclopedia of Current Information about Each Branch of the Electrical Industry, with a Dictionary of Electrical Terms and a Classified Directory of Electrical and Related Products and Their Manufacturers in the United States and Canada* (Chicago: Electrical Trade Publishing Company, 1923), 563. "Walter Lemmon, Inventor, Is Dead," *New York Times*, 21 March 1967, p. 46; Donald H. Short, "Woodrow Wilson First 'DX' Broadcaster: New York Radio Man Recalls Incidents of 1919," *Times Leader Evening News* (Wilkes Barre, Pennsylvania), 16 February 1924, p. 4; "Radio Programs-Development-Boston," RF 35118, File 3513, Box 295, Rockefeller Foundation Project Records. Quotes pertaining to Lemmon's vision of the WWBF are from "Story of W1AXL"; "Harvard's Classes to Be on World Radio," *New York Times*, 21 February 1937, p. 21; and "A New International Radio Service for Broadcasting Cosmic Data," *Scientific Monthly* 44.3 (1937), 288. For Depression-era radio and its profitability, see Craig, *Fireside Politics*, 14–17.

9. "Radio Programs-Development-Boston," RF 35118, File 3513, Box 295, Rockefeller Foundation Project Records. Hilmes, *Network Nations*, 115–16; Berg, *On the Shortwaves*, 47–59.

10. "Radio Programs-Development-Boston," RF 35118, File 3513, Box 295, Rockefeller Foundation Project Records.

11. Peter Dobkin Hall, "Business, Philanthropy, and Education in the United States," *Theory into Practice* 33.4 (1994), 212–13; Anne-Emanuelle Birn, "Public Health or Public Menace? The Rockefeller Foundation and Public Health in Mexico, 1920–1950," *Voluntas: International Journal of Voluntary and Nonprofit Organizations* 7.1 (1996), 37–39; Ann Zulwaski, *Unequal Cures: Public Health and Political Change in Bolivia, 1900–1950* (Durham, NC: Duke University Press, 2007), 86; Ann Vogel, "Who's Making Global Civil Society: Philanthropy and US Empire in World Society," *British Journal of Sociology* 57.4

(2006), 635–55; Jeffrey D. Brison, *Rockefeller, Carnegie, and Canada American Philanthropy and the Arts and Letters in Canada* (Montréal: McGill-Queen's University Press, 2005); Rockefeller Foundation Annual Reports from the years 1934–1940, https://www.rockefellerfoundation.org/annual-reports. On American exceptionalism and US foreign relations, see Emily Rosenberg, *Spreading the American Dream: American Economic and Cultural Expansion, 1890–1945* (New York: Hill and Wang, 1982). Rosenberg addresses the Rockefeller Foundation specifically on pp. 119–20; see also Michael Adas, *Dominance by Design: Technological Imperatives and America's Civilizing Mission* (Cambridge, MA: Belknap Press of Harvard University Press, 2006), 210–12.

12. "Progress Report of 1950 Operations," Folder 1192, Box 135, Messrs. Rockefeller Cultural Interests Records. "Foundation Broadcasts Messages of Hope to Oppressed People Abroad," *New York World-Telegram*, 30 January 1941, copy of article in Folder 3533, Box 296, Rockefeller Foundation Project Records. Although from 1950, the quoted portion of the "Progress Reports" taken from p. 7 speaks to the more enduring aspects of Lemmon's vision and looks back at the prewar context into which the WWBF broadcast.

13. Approved Motion RF 35116, "Radio Programs-Development-Boston," 21 June 1935; "Request presented to the Humanities and circulated to the officers for discussion in conference Thursday," 1935 (no specific day or month recorded on this internal memo); Approved Motion 36051, "World Wide Broadcasting Foundation—Station W1XAL-Boston," 15 April 1936; Approved Motion RF 38056, "World Wide Broadcasting Foundation," 6 April 1938; Folder 3513, Box 295, Rockefeller Foundation Project Records.

14. For the WWBF-authored history, see "Story of W1XAL," 6 May 1935, Folder 3513, Box 295, Rockefeller Foundation Project Records. Lemmon quoted in "Progress Report of 1950 Operations," p. 8, Folder 1192, Box 135, Messrs. Rockefeller Cultural Interests Records. Though the document was written in 1950, the section speaking to language education harkens back to the prewar initiatives discussed here.

15. Paul F. Lazarsfeld, *Radio and the Printed Page: An Introduction to the Study of Radio and Its Role in the Communication of Ideas* (New York: Duell, Sloan, and Pierce, 1940). Lazarsfeld's conclusions are quoted in Patrick R. Parsons, "The Lost Doctrine: *Suggestion Theory* in Early Media Effects Research," *Journalism and Communications Monographs* 23.2 (2021), 119. The enduring significance of this book to radio studies, its connection to the Rockefeller research, and Lazarfeld's early social constructivist inclinations are addressed in Michael Stamm, "Paul Lazarsfeld's *Radio and the Printed Page*: A Critical Reappraisal," *American Journalism* 27.4 (2010), esp. 50–51, 54. See also Horten, *Radio Goes to War*, 25–27.

16. Lazarsfeld to Marshall, 11 March 1938, Folder 3520, and "Summary of Opinion of W1XAL Situation," undated (received 19 March 1938), Folder 3521, Box 295, Rockefeller Foundation Project Records (quotes are from the "Summary" document).

17. Rodney Koeneke, *Empires of the Mind: I. A. Richards and Basic English in China* (Stanford, CA: Stanford University Press, 2004), 4; George Watson, "The Amiable Heretic: I. A. Richards, 1893–1979," *Sewanee Review* 104.2 (1996) 254; C. K. Ogden, "Will Basic English Become the Second Language," *Public Opinion Quarterly* 8.1 (1944), 3–9 (Ogden

quote is from p. 5); Richards quoted in I. A. Richards, "Basic English and Its Applications," *Journal of the Royal Society of the Arts* 87.4515 (1939) 747, and I. A. Richards, *Basic English and Its Uses* (New York: W. W. Norton, 1943), 131–32.

18. The verb count comes to eighteen if you include the two auxiliaries *will* and *may*. Richards, "Basic English and Its Applications," 739–41, 749; Stuart Brown, "Basic English," Theresa Enos, ed., *Encyclopedia of Rhetoric and Composition: Communication from Ancient Times to the Information Age* (New York: Routledge, 1996), 68; Koeneke, *Empires of the Mind*, 3–4.

19. Emily Rosenberg, "Transnational Currents in a Shrinking World," Rosenberg, ed., *A World Connecting, 1870–1945* (Cambridge, MA: Belknap Press of Harvard University Press, 2012), 850–55.

20. The Rockefeller Foundation's rationale for supporting the Basic English initiative in China and Japan is quoted from *Rockefeller Foundation Annual Report 1933*, available at https://www.rockefellerfoundation.org/wp-content/uploads/Annual-Report-1933-1.pdf. Richards, "Basic English and Its Applications" (Richards quoted on pp. 738, 746; Paget quoted on p. 750; unnamed individual quoted on p. 751). Background information on Sir Richard Arthur Surtees Paget is from Harry Lowery and John Bosnell, "Paget, Sir Richard Arthur Surtees, second baronet," *Oxford Dictionary of National Biography*, http://www.oxforddnb.com/view/article/35358 (2008). Lemmon quoted in Thomas Kellend, "Foundation Broadcasts Messages of Hope to Oppressed Peoples Abroad," *New York World-Telegraph*, 30 January 1941, clipping in Folder 3533, Box 296, Rockefeller Foundation Project Records.

21. Stuart Anderson, *Race and Rapprochement: Anglo-Saxonism and Anglo-American Relations, 1895–1904* (Rutherford, NJ: Fairleigh Dickinson University Press, 1981); Reginald Horsman, *Race and Manifest Destiny: The Origins of American Racial Anglo-Saxonism* (Cambridge, MA: Harvard University Press, 1981); David Reynolds, "Rethinking Anglo-American Relations," *International Affairs* 65.1 (1988/1989), esp. 95; Sasha Abramsky, "The Anglo-American Misalliance," *World Policy Journal* 25.1 (2008), esp. 75; Keith Neilson, "Perception and Posture in Anglo-American Relations: The Legacy of the Simon-Stimson Affair, 1932–1941," *International History Review* 29.2 (2007), esp. 336; Robert Self, "Perception and Posture in Anglo-American Relations: The War Debt Controversy in the 'Official Mind,' 1919–1940," *International History Review* 29.2 (2007), esp. 282–83; Dianne Kirby, "Divinely Sanctioned: The Anglo-American Cold War Alliance and the Defence of Western Civilization and Christianity, 1945–48," *Journal of Contemporary History* 35.3 (2000), esp. 387; Steve Marsh, "The Special Relationship and the Anglo-Iranian Oil Crisis, 1950–4," *Review of International Studies* 24.4 (1998), esp. 538.

22. Richards, "Basic English and Its Applications," 747; Churchill to Roosevelt, 20 April 1944, and Roosevelt to Hull, 5 June 1944, Folder "Great Britain: Winston Churchill, 1944–45," Box 37, Series I: Safe Files, The President's Secretary's File, 1933–1945, Franklin Roosevelt Presidential Library, Hyde Park, New York, http://www.fdrlibrary.marist.edu/_resources/images/psf/psfa0500.pdf, pp. 21–26. Ogden, "Will Basic English Become the Second Language," 4.

23. Grant-in-Aid request for funding Pin Pin T'an's WWBF work (RA M 253), approved 1 December 1939, memorandum noting scheduled meeting between T'an

and Richards, 23 October 1939, Richards to RF Humanities Division Director David H. Stevens, 26 October 1939, Folder 3532, Box 296, Rockefeller Foundation Project Records. "I. A. Richards, Author, Teacher and Literary Critic, Is Dead at 86," *New York Times*, 8 September 1979, pp. 1, 36; Koeneke, *Empires of the Mind*, 5–6.

24. T'an, "Report: Basic English Broadcasts over Station WRUL, September 16, 1939 through July 12, 1940" (direct quote about Basic as an "experiment" on p. 1 of the report) and T'an, "Basic English—A World Secondary Language," *World Wide Listener*, November 1939, pp. 8–9 (direct quote on "language habits" from p. 8), Folder 3534; memorandum of conversation with Charlotte Tyler regarding T'an (contains direct quote praising T'an's performance), 8 November 1939, Folder 3532; Box 296, Rockefeller Foundation Project Records.

25. T'an, "Report: Basic English Broadcasts over Station WRUL, September 16, 1939 through July 12, 1940" (T'an's quote noting de Wolf's support is from p. 2), and T'an, "Basic English—A World Secondary Language," *World Wide Listener*, November 1939, 8–9 (includes the direct quotes of T'an effusing about the benefits of Basic English), Folder 3534, Box 296, Rockefeller Foundation Project Records.

26. T'an, "Report: Basic English Broadcasts over Station WRUL, September 16, 1939 through July 12, 1940" (T'an's quotes are from p. 2), Folder 3534, and Tyler to Stevens, 20 June 1940, Folder 3533, Box 296, Rockefeller Foundation Project Records.

27. Tyler to Stevens, 20 June 1940, Folder 3533, and T'an, "Report: Basic English Broadcasts over Station WRUL, September 16, 1939 through July 12, 1940," p. 2, Folder 3534, Box 296, Rockefeller Foundation Project Records.

28. T'an, "Report: Basic English Broadcasts over Station WRUL, September 16, 1939 through July 12, 1940," pp. 3–4, Folder 3534, Box 296, Rockefeller Foundation Project Records.

29. T'an, "Report: Basic English Broadcasts over Station WRUL, September 16, 1939 through July 12."

30. Marshall to C. J. Friedrich (Graduate School of Public Administration, Harvard University), 6 July 1940, Folder 3525; Ogden to T'an, 21 June 1940, reprinted in T'an, "Report: Basic English Broadcasts over Station WRUL, September 16, 1939 through July 12, 1940" (see p. 4 in the section of the report entitled "Excerpts from Letters"), Folder 3534; and Grant-in-Aid Request to continue WWBF Basic English Funding, 21 August 1940, Folder 3532, Box 296, Rockefeller Foundation Project Records.

31. Grant-in-Aid Request to continue WWBF Basic English Funding, 21 August 1940, Folder 3532, and memorandum of conversation between Marshall and T'an, 27 February 1941, Folder 3533, Box 296, Rockefeller Foundation Project Records. "Progress Report of 1950 Operations," Folder 1192, Box 135, Messrs. Rockefeller Cultural Interests Records.

32. Memorandum by Marshall recording his observations following two days of meetings at W1AXL, 28 and 29 December 1936, Folder 3516, Box 295, and John Marshall to C. J. Friedrich (Graduate School of Public Administration, Harvard University), 6 July 1940, Folder 3525, Box 296, Rockefeller Foundation Project Records.

33. Lemmon to George J. Beal (Comptroller, Rockefeller Foundation), 24 October 1940, Folder 3533, and memorandum of conversation between Marshall and T'an, 27

February 1941, Folder 3533, Box 296, Rockefeller Foundation Project Records. Lemmon to Mr. Judd (Manager, Boston Symphony Orchestra), 4 November 1940, Folder 1192, Box 135, Messrs. Rockefeller Cultural Interests Records.

34. Rafael Mojica (Dominican Republic) to Station WRUL, undated, translations of letters received by the WWBF from July 1940 through June 1941, Survey of Basic Correspondence from Latin America (included with sample broadcast transcript entitled "Un Idioma Internaciónal Por Una Estación Internaciónal"); J. Antonio Maya J. (Nicaragua) to the WWBF, 25 June 1940, Gladys Heider Rivero (Cuba) to the WWBF, 6 August 1940, and Carlos Aquila (Havana, Cuba) to the WWBF, 7 June 1940 (included with Walter Lemmon to Rockefeller Foundation Humanities Director David H. Stevens, 13 August 1940); Folder 3533, Box 296, Rockefeller Foundation Project Records.

35. C. Flores (Argentina) to the WWBF, 16 June 1940 (included with Lemmon to Stevens, 13 August 1940); Edgar Salas B. (Costa Rica) to the WWBF, undated, translations of letters received by the WWBF from July 1940 through June 1941, Survey of Basic Correspondence from Latin America (included with "Un Idioma Internaciónal" transcript); Folder 3533, Box 296, Rockefeller Foundation Project Records.

36. Bruce Lenthall, *Radio's America: The Great Depression and the Rise of Modern Mass Culture* (Chicago: University of Chicago, 2007), 97 (includes quote about Fireside Chats); see also Loviglio, *Radio's Intimate Public*.

37. Michael Salwen, *Radio and Television in Cuba: The Pre-Castro Era* (Ames: Iowa State University Press, 1994), 60; Matthew B. Karush, *Culture of Class: Radio and Cinema in the Making of a Divided Argentina, 1920–1946* (Durham, NC: Duke University Press, 2012), 52, 66; Joy Elizabeth Hayes, *Radio Nation: Communication, Popular Culture, and Nationalism in Mexico, 1920–1950* (Tucson: University of Arizona Press, 2000), 22, 60 (latter page includes the direct quote from Mexico).

38. John E. Joseph, *Language and Identity: National, Ethnic, Religious* (New York: Palgrave Macmillan, 2004) 70–71; Eidsheim, "Marion Anderson and 'Sonic Blackness' in American Opera," 644–45. Domingo Rios to WWBF, undated, translations of letters received by the WWBF from July 1940 through June 1941, Survey of Basic Correspondence from Latin America (included with "Un Idioma Internaciónal" transcript), Folder 3533, Box 296, Rockefeller Foundation Project Records.

39. Regina Cortina, "Introduction," Regina Cortina, ed. *The Education of Indigenous Citizens* (Tonawanda, NY: Multilingual Matters, 2014), 1–2, 18; Luis Enrique Lopez, "Indigenous Intercultural Bilingual Education in Latin America: Widening Gaps between Policy and Practice," Cortina, ed., *The Education of Indigenous Citizens*, 19–49; Mary Kay Vaughan, *Cultural Politics in Revolution: Teachers, Peasants, and Schools in Mexico, 1930–1940* (Tucson: University of Arizona Press, 1997), 114–15, 126; Yamilet Hernandez-Galano and Lyding R. Rodriquez Fuentes, "Schooling in Cuba," 156–57, 171, note 63, Javier Sáenz Obregon and Oscar Saldarriaga Vélez, "Schooling in Colombia," 105, Silvina Gvirtz, Jason Beech, and Angela Oria, "Schooling in Argentina," 12–13, and Rolando Poblete Melis, "Schooling in Chile," 87, in Silvina Gvirtz and Jason Beech,

eds., *Going to School in Latin America* (Westport, CT: Greenwood Press, 2008); Charles A. Howard, "Electrification for Latin America," *Electrical Engineering* 63.2 (1944), 43–46.

40. Barbara Seidelhofer, *Understanding English as a Lingua Franca* (New York: Oxford University Press, 2011), 168–70.

41. Seidelhofer, *Understanding English as a Lingua Franca*, 166; Ogden, "Will Basic English Become the Second Language," 7, 9 (direct quote is from p. 7).

42. Alan McPherson, "*Antiyanquismo*: Nascent Scholarship, Ancient Sentiments," McPherson, ed., *Anti-Americanism in Latin America and the Caribbean* (New York: Berghahn Books, 2006), 14–19; and Alan McPherson, "Anti-Americanism in Latin America," Brendon O'Connor, ed., *Anti-Americanism: History, Causes, and Themes*, vol. 3 (Westport, CT: Greenwood World Publishing, 2007), 86.

43. I. A. Richards and Christine Gibson, "Learning Basic English: An Overall View," *English Journal* 34.6 (1945), 305.

44. Sherman Cochran, *Big Business in China: Sino-Foreign Rivalry in the Cigarette Industry, 1890–1930* (Cambridge, MA: Harvard University Press, 1980); Watson, "The Amiable Heretic," 257.

45. See esp. Tyler to Marshall, 3 August 1940, Marshall to Tyler, 7 August 1940, E. E. Salisbury, Assistant to the Commissioner, US Immigration and Naturalization Service to Pin Pin T'an, 19 December 1940, Tyler to Salisbury, 15 January 1941 (letter is dated in error as written in 1940), and memorandum of interview with T'an by Marshall, 27 February 1941, Folder 3533, Box 296, Rockefeller Foundation Project Records; "Romance Started in Pre War Peipin Culminates in New York," *News of China: United China Relief*, January 1945, p. 6.

46. Hilmes, *Network Nations*, 118–19; Michele Hilmes, "The New Vehicle of Nationalism: Radio Goes to War," Jonathan Auerbach and Russ Castronovo, eds., *The Oxford Handbook of Propaganda Studies* (New York: Oxford University Press, 2014), 202, 212–14; Nicholas Cull, *Selling War: The British Propaganda Campaign Against American "Neutrality" in World War II* (New York: Oxford University Press, 1995), 133; Michael B. Salwen, "Lemmon, Walter S.," Guido Hermann Stempel and Jacqueline Nash Gifford, eds., *Historical Dictionary of Political Communication* (Westport, CT: Greenwood Press, 1999), 79. The "weapon for peace" quote is from a 1952 promotional pamphlet, "WRUL. . . . The Direct Radio Link to Business Activities in Every Latin American Market," Folder 1189, Box 135, Messrs. Rockefeller Cultural Interests Records.

Chapter 5. "A Workable Scheme to Quiet the Panaman Clamor"

1. "Our Isolation," editorial from the *Star and Herald* (Panama City, Panama), 26 June 1925, enclosed with American Minister John Glover South to Secretary of State Frank B. Kellogg, 26 June 1925, 819.74/134, DSNA 1910–29.

2. Raymond Leslie Buell, "Union or Disunion in Central America?," *Foreign Affairs* 11.3 (1933), 478.

3. Alfred Thayer Mahan, *The Interest of America in Sea Power, Present and Future* (New York: Little, Brown, 1897), 260. Ira Bennet, *History of the Panama Canal: Its Construction and Builders* (Washington, DC: Historical Publishing Company, 1915), 322. Edmonds

was quoted in an editorial piece he authored, "World Influence of the Panama Canal Is Topic of Editor R. H. Edmonds," *Daily Herald* (Gulfport and Biloxi, Mississippi), 31 January 1914, p. 8; Yoshimitsu Ide, "The Significance of Richard Hathaway Edmonds and His *Manufacturers' Record* in the New South," (PhD dissertation: University of Florida, 1959), esp. 19.

4. This assessment of frontpage coverage is based on news coverage on August 14 and 15, 1914, via newspapers.com. The *San Francisco Chronicle* and *Los Angeles Times* were two notable exceptions and found space on the front page to tout the Canal's opening alongside war news. The ports in both cities stood to benefit greatly from the reduced distance and time it took to reach the East Coast.

5. "History of Wireless Telegraph Agreement Between Panama and the United States," Memorandum of the Division of Latin American Affairs, 15 May 1914, 819.74/91, DSNA 1910–29.

6. Rosenberg, *Spreading the American Dream*, esp. 59, 230–31; Schwoch, *The American Radio Industry and Its Latin American Activities*, esp. 22–23; Aitken, *The Continuous Wave*, 338; Jonathan Reed Winkler, *NEXUS: Strategic Communications and American Security in World War I* (Cambridge, MA: Harvard University Press, 2008), 183–84; Charles Morrow Wilson, *Empire in Green and Gold: The Story of the American Banana Trade* (New York: Holt, Reinhart, and Winston, 1947), 154–55; "GE to Install Six High Power Tube Transmitters," *Radio News*, January 1924, 1010.

7. Walter LaFeber, *The Panama Canal: The Crisis in Historical Perspective*, updated edition (New York: Oxford University Press, 1989), 18–19; Julia Greene, *The Canal Builders: The Making of America's Empire at the Panama Canal* (New York: Penguin Press, 2009), 6, 21–22, 305. For radio's appeal to the Panamanian government, see Panama's Secretary of Foreign Affairs Ernesto Lefevre to the American Minister in Panama William Jennings Price, 14 August 1914, Enclosure 2 in dispatch No. 268, 819.74/58, DSNA 1910–29.

8. Acting Secretary of the Navy Beekman Winthrop to the Secretary of State Philander C. Knox, 13 December 1911, 819.74/3 (includes note from the Division of Latin American Affairs Chief William T. S. Doyle directed to Assistant Secretary of State Huntington Wilson, 16 December 1911); Department of State Counselor J. B. Moore to the Secretaries of War and Navy, 29 August 1913, 819.74/30; and Daniels to Bryan, 22 November 1913, 819.74/35; DSNA 1910–29. David McCreery, "Wireless Empire: The United States and Radio Communications in Central America and the Caribbean, 1904–1926," *South Eastern Latin Americanist* 37 (1993), 24. Rita Zajácz, "Fragmented Imperialism: US Control over Radio in Panama, 1914–36," *International Communications Gazette* 74.1 (2012), 82–83.

9. United States, Library of Congress, *Background Documents Relating to the Panama Canal* (Washington, DC: US Government Printing Office, 1977), esp. 280–81. American Minister Washington Dodge to Secretary of State Philander C. Knox, 14 February 1912, 819.74/12; "Operation of Wireless Telegraph Stations on the Isthmus of Panama," 4 April 1912, 819.74/13; Secretary of Navy Josephus Daniels to Secretary of State William Jennings Bryan, 22 November 1913, 819.74/35; "History of Wireless Telegraph

Agreement Between Panama and the United States," Memorandum of the Division of Latin American Affairs, 15 May 1914, 819.74/91; DSNA 1910–29.

10. "History of Wireless Telegraph Agreement Between Panama and the United States," p. 7, 14 May 1914, 819.74/91, DSNA 1910–29; J. Michael Hogan, *The Panama Canal in American Politics: Domestic Advocacy and the Evolution of Policy* (Carbondale: Southern Illinois University Press, 1986), 64–65; Greene, *Canal Builders*, 320; Robert Harding, "The Military Foundations of Panamanian Politics: From the National Police to the PRD and Beyond" (PhD dissertation: University of Miami, 1998), 1.

11. Micah Wright, "Unilateral Pan-Americanism: Wilsonianism and the American Occupation of Chiriquí," *Diplomacy and State Craft* 26.1 (2015), esp. 48–49. Porras is quoted in John Biesanz and Mavis Biesanz, *The People of Panama* (New York: Columbia University Press, 1955), 96.

12. Lefevre to American Minister William Jennings Price, enclosed with Price to Bryan, 14 August 1914, 819.74/58, DSNA 1910–29.

13. LaFeber, *The Panama Canal*, 50–56; Frank Schumacher, "Embedded Empire: The United States and Colonialism," *Journal of Modern European History* 14.2 (2016), 206–7; Michael Baud, "Imagining the Other: Michael Taussig on Mimesis, Colonialism, and Identity," *Critique of Anthropology* 17.1 (1997), 105.

14. Price to Bryan, 22 August 1914, 819.74/60 (enclosures include the original presidential decree rejected by the United States and Minister Price's objections to its temporary nature); Price to Bryan, 23 August 1914; Bryan to the American Legation, 25 August 1914, 819.74/57; Lefevre to Price, enclosed with Price to Bryan, 14 August 1914, 819.74/58; Price to Bryan, 27 August 1914, 819.74/61; Porras, Presidential Decree 130 of 1914, enclosed with Price to Bryan, 29 August 1914, 819.74/62; DSNA 1910–29. "Panama President, Former Student at Pennsylvania U," 18 February 1920, *Spokane Daily Chronicle* (Spokane, Washington), p. 13 (this story was nationally syndicated); "Newspaper Work in Canal Zone Includes Reporting Bull Fights and Tea Parties," *Hartford Courant* (Hartford, Connecticut), 10 October 1920, p. 1. Rita Zajácz, "Fragmented Imperialism," 84; Enrique Lefevre, *Más allá del olvido: Ernesto Lefevre y el imperialismo yanky* [also published under the title *Cuidadano Ejemplar*] (Panama City: Imprenta Universidad de Panama, 1972), 72–73; Thomas Leonard, *Historical Dictionary of Panama* (Lanham, MD: Rowman and Littlefield, 2015), 177.

15. The Punto Mala site had previously been designated as "property" of the Panama Canal, thus the United States already held sovereign rights, whereas the La Palma and Puerto Obaldía stations were built on sites that had otherwise fallen under Panama's jurisdiction. "The Radio System on the Isthmus," *Panama Canal Record*, 5 October 1921, 130–31. Daniels to Secretary of State Robert Lansing, 7 November 1918, 819.74/80; Price to Lansing, 20 December 1918, 819.74/82; Commandant of the 15th Naval District L. R. Sargent to Chargé d'Affaires Elbridge Gerry Green, 12 June 1918, enclosed with Price to Lansing, 20 December 1918, 819.74/82; "Status of Wireless Stations Being Erected in Panama by the United States Navy Department," Memorandum from the Department of State Office of the Solicitor, 14 December 1918, 819.74/81; Polk to Daniels, 9 January 1919, 819.74/80; DSNA 1910–29.

16. Price to the Fiscal Agent for Panama A. T. Ruan, 23 January 1920 (enclosure 3) and Price to Lefevre, 23 January (enclosure 4), included with Price to Lansing, 26 January 1920, 819.74/88, and Assistant Secretary in Charge of Panama's Foreign Office Evenor Hazera to Price, 12 February 1920 (enclosure 2), included with Price to Secretary of State Bainbridge Colby, 8 April 1920, 819.74/89, DSNA 1910–29.

17. John Major, *Prize Possession: The United States and the Panama Canal, 1903–1979* (New York: Cambridge University Press, 1993), 139–41, and George W. Baker Jr., "The Wilson Administration and Panama, 1913–1921," *Journal of InterAmerican Studies* 8.2 (1966), 292; Greene, *The Canal Builders*, 329–32; Wright, "Unilateral Pan-Americanism," esp. 47–48.

18. Price to Colby, 15 October 1920, esp. enclosure 2, "National Assembly—Session of October 9th," 9 October 1920, and enclosure 5 "IV—Report of the Committee," 12 October 1920 (both enclosures are typed translated reproductions that were originally published in Spanish in the 8 October and 12 October editions of the Canal Zone's *Star and Herald* newspaper), 819.74/93, DSNA 1910–29. See also Major, *Prize Possession*, 139.

19. Lefevre's address to the National Assembly in October 1920 is reprinted in Lefevre, *Más allá del olvido*, 221–25. The direct quotes, which have been translated from their original Spanish into English, are from pp. 221 and 224–25. See also Price to Colby, 15 October 1920, enclosure 7, "A Famous Document Read Day Before Yesterday by Mr. E. T. Lefevre Before the Assembly," 13 October 1920 (enclosure is a typed translated reproduction originally published in Spanish in the 13 October edition of the Canal Zone's *Star and Herald* newspaper), 819.74/93, DSNA 1910–29. "Ernesto Lefevre Dead: Ex-President of Panama Dies After an Operation," *New York Times*, 26 December 1922, p. 12.

20. Price to Colby, 15 October 1920, 819.74/93 (includes reaction to Lefevre's speech from Spanish language section of the 12 October 1920 edition of the *Star and Herald*); Price to Secretary of State Charles Evans Hughes, 23 June 1921, 819.74/95; Dana Gardner Munro of the Division of Latin American Affairs to Assistant Secretary of State Leland Harrison, 30 March 1922, 819.74/103; DSNA 1910–29.

21. Garay to South, 14 August 1922 and South to Garay, 13 September 1922, enclosed with South to Hughes, 24 January 1923, 819.74/112, DSNA 1910–29.

22. Canal Zone Governor Jay J. Morrow to Chief of the State Department's Office of the Panama Canal, 9 September 1922; for the Panama Radio Club, see Chairman Richard D. Prescott and Secretary-Treasurer A. J. Lind of the Panama Radio Club to Morrow, 29 November 1922; both documents enclosed with Secretary of War John Weeks to Secretary of State Charles Evans Hughes, 11 January 1923, 819.74/108, DSNA 1910–29. For broadcasting's World War I origins, see Douglas, *Inventing American Broadcasting*, esp. 276, 293, 297, and Hilmes, *Network Nations*, esp. 34, 37. For broadcasting's development in Latin America and the Caribbean, see Schwoch, *The American Radio Industry and Its Latin American Activities*; Bronfman, *Isles of Noise*; Castro, *Radio in Revolution*; Hayes, *Radio Nation*; Karush, *Culture of Class*.

23. Alfaro to Hughes, 19 December 1922, 819.74/106, DSNA 1910–29.

24. Morrow to Chief of the State Department's Office of the Panama Canal, 9 Sep-

tember 1922, and Morrow, "Radio Broadcasting in or adjacent to the Canal Zone," Memorandum for the Secretary of War, 8 January 1923, both documents enclosed with Secretary of War John Weeks to Hughes, 11 January 1923, 819.74/108; Denby to Hughes, 23 January 1923, 819.74/109; DSNA 1910–29. J. Farrell, "The Nerve System of the Sea," *Radio News*, November 1923, 522, 618–19, and "Broadcasting in Panama," *World Wide Wireless*, January 1924, 45.

25. "International Radio Notes," *Radio News*, May 1924, 1673. South to Hughes, 24 January 1923, 819.74/110; South to Hughes, 19 February 1923, 819.74/113; "Contra Memorandum," enclosed with South to Hughes, 30 April 1923, 819.74/115; Hughes to American Chargé d'Affaires ad interim Edward L. Reed, enclosed with Under Secretary of State William Phillips to Weeks, 19 July 1923, 819.74/117; DSNA 1910–29.

26. Hughes to Denby, 19 July 1923, 819.74/117, DSNA 1910–29.

27. Munro to Harrison, 30 March 1922, 819.74/103, DSNA 1910–29.

28. United States, Library of Congress, *A List of American Doctoral Dissertations Printed in 1918* (Washington, DC: US Government Printing Office, 1921), 118; Dana Gardner Munro, *A Student in Central America* (New Orleans: Tulane University Press, 1983), ix, xi, 4; Alfonso A. Narvaez, "Dana G. Munro, 97, A Retired Professor and Ex-Diplomat," *New York Times*, 19 June 1990, p. B6. For "tactless" quote, see Major, *Prize Possession*, 149. For "the other end of the cable line" quote, see Dana Gardner Munro's review of William L. Beaulac, *Career Ambassador* (New York: The MacMillan Company, 1951), *Hispanic American Historical Review* 32.1 (1952), 116.

29. Max Paul Friedman, "The Good Neighbor Policy," *Oxford Research Encyclopedia of Latin American History* (2018), https://doi.org/10.1093/acrefore/9780199366439.013.222. See also Alexander DeConde, *Herbert Hoover's Latin American Policy* (Palo Alto, CA: Stanford University Press, 1951); Alan McPherson, "Herbert Hoover, Occupation Withdrawal, and the Good Neighbor Policy," *Presidential Studies Quarterly* 44.4 (2014), 623–39; Bryce Wood, *The Making of the Good Neighbor Policy* (New York: Columbia University Press, 1961); Irwin F. Gellman, *Good Neighbor Diplomacy: United States Policies in Latin America, 1933–1945* (Baltimore: Johns Hopkins University Press, 1979); Dana Gardner Munro, *The United States and the Caribbean Republics, 1921–1933* (Princeton, NJ: Princeton University Press, 1974), esp. 272–73; Dana G. Munro, *Intervention and Dollar Diplomacy in the Caribbean, 1900–1921* (Princeton, NJ: Princeton University Press, 1964), viii (contains quote on previous policy experience).

30. Major, *Prize Possession*, 111, 144 (contains White's quotes). Gellman, *Good Neighbor Diplomacy*, 3; Friedman, "The Good Neighbor Policy." Dawson to Chief of the Division of Latin American Affairs Walter C. Thurston, 14 February 1931, 819.74/189, and South to Secretary of State Frank Kellogg, 20 August 1927, 819.6156T61/16, DSNA 1910–29.

31. Lawrence O. Ealy, *The Republic of Panama in World Affairs, 1903–1950* (Philadelphia: University of Pennsylvania Press, 1951), 63–64; Leonard, *Historical Dictionary of Panama*, 25.

32. Major, *Prize Possession*, 112–13, 183–84; Thomas Leonard, "The United States and Panama: Negotiating the Aborted 1926 Treaty," *Mid America: An Historical Review* 61.2 (1979).

33. South to Kellogg, 17 February 1927, 819.6156T61/5, DSNA 1910–29 (included with this correspondence is a copy of the contract). The railroad concession of this new agreement permitting the construction of a new railroad line out of Tonosí was also a direct challenge to the US monopoly over railroads in Panama. This controversial claim was based on a nineteenth-century treaty signed when Colombia ruled the isthmus.

34. Jason M. Colby, *The Business of Empire: United Fruit, Race, and U. S. Expansion in Central America* (Ithaca, NY: Cornell University Press, 2011), 90–91, 121; Philippe Bourgois, "One Hundred Years of United Fruit Company Letters," Steve Striffler and Mark Moberg, eds., *Banana Wars: Power Production, and History in the Americas* (Durham, NC: Duke University Press, 2003), 117; James W. Martin, *Banana Cowboys: The United Fruit Company and the Culture of Corporate Colonialism* (Albuquerque: University of New Mexico Press, 2018), esp. 6, 13, 64–66, 73.

35. South to Kellogg, 17 February 1927, 819.6156T61/5 and South to Kellogg, 11 March 1927, 819.6156T61/8, DSNA 1910–29.

36. President of Tropical Radio and Telegraph Company Victor M. Cutter to Secretary of War Dwight F. Davis, 16 June 1929, 819.74/152, and Day to the Governor of the Panama Canal, 14 February 1927, enclosed with South to Kellogg, 17 February 1927, 819.6156T61/5, DSNA 1910–29.

37. Schumacher, "Embedded Empire," esp. 224; LaFeber, *The Panama Canal*, 52–53 (esp. n. 8); Ronan Van Rossem, "The World System Paradigm as General Theory of Development: A Cross-National Test," *American Sociological Review* 61.3 (1996), esp. 519–24; George A. Barnett, Thomas Jacobson, Young Choi, and Sulien Sun-Miller, "An Examination of the International Telecommunication Network," *Journal of International Communication* 3.2 (1996), 39.

38. State Department approval of the contract documented in Secretary of State Henry Stimson to the Tropical Radio Telegraph Company, 25 November 1929, 819.74/153, DSNA 1910–29.

39. American Chargé d'Affaires Benjamin Muse to Kellogg, 17 June 1929, 819.6341P19/33, DSNA 1910–29.

40. Dawson to Thurston, "American control of radio communications in Panama," 5 January 1931, 819.74/195, and American Minister Roy T. Davis to Stimson, 21 October 1932, 819.74/238, DSNA 1930–39.

41. Unsigned letter from the State Department's Treaty Division to Dawson, 21 July 1931, 819.74/206; Dawson to Thurston and White, Memorandum, 1 April 1931, 819.74/204; and William R. Vallance, "Memorandum of Conference with Commander Lammers Concerning American Control of Radio in Panama," 20 June 1931, 819.74/204, DSNA 1930–39.

42. President of All America Cables John L. Merrill to Stimson, 12 September 1931, 819.74 All America Cables Co./18, DSNA 1930–39.

43. Memorandum of conversations between Merrill and Navy Commander Howard M. Lammers, 1 October 1931, 819.74 All America Cables Co./28; Chief of Naval

Operations to the Secretary of the Navy, 28 September 1931, enclosed with American Chargé d'Affaires Howard Bucknell Jr. to Stimson, 19 November 1931, 819.74/224; Federal Radio Commission Chairperson McKenzie Saltzman to Stimson, 4 November 1931, 819.74 All America Cables Co./30; White, Memorandum of Conversation with International Telephone and Telegraph Vice President Frank Page, 7 July 1932, 819.74 All America Cables Co./57 (contains White's quotes); White to Stimson, 24 March 1932, 819.74 All America Cables Co./45; Under Secretary of State William R. Castle, "Memorandum of Conversation with Secretary of the Navy," 21 December 1932, 819.74 All America Cables Co./86; Chief of the Division of Latin American Affairs Edwin Wilson to White, handwritten note, 19 March 1932, and undated fifteen-page draft of a letter composed for the secretary of state's signature and responding to Adams's 12 November 1931 letter, 819.74 All America Cables Co./36; Stimson to Adams, 25 November 1932, 819.74 All America Cables Co./60 and Stimson to All America Cables, 7 February 1933, 819.74 All America Cables Co./75; Merrill to White, 30 March 1933, 819.74 All America Cables Co./93; DSNA 1930–39.

44. Arosemena's and Stimson's quotes are from an unsigned memo outlining the full program celebrating the inauguration of the new radio link, 23 February 1933, 819.76 Tropical Radio Telegraph Co./37, DSNA 1930–39.

45. Leonard, "Negotiating the Aborted 1926 Treaty," 199.

46. Davis to Secretary of State Cordell Hull, "Control of Radio in Panama," 10 September 1933, 819.74/253; "Conference with President Arias in the Secretary's Office on the Morning of October 9, 1933," Memorandum, Division of Latin American Affairs, 10 October 1933, 819.74/257 (contains Arias's "indignity" quote); "Memorandum of Points Agreed to By President Roosevelt and President Arias," enclosed with Assistant Secretary of State Jefferson Caffery to American Minister Antonio Gonzalez, 20 October 1933, 711.19/175; "Radio Board," Memorandum of conversation on 14 November 1933 between American Minister Antonio Gonzalez and President Arias from the American Legation in Panama, enclosed with Gonzalez to Hull, 17 November 1933, 819.74/262; President Franklin D. Roosevelt to Acting Secretary of State, 19 December 1933, 819.74/263; DSNA 1930–39. See also Major, *Prized Possession*, 287–88.

47. "Radio in Panama," 13 March 1934, 819.76 Tropical Radio Company/69, and Hull to Arosemena, 12 April 1934, 819.74/279, DSNA 1930–39.

48. "Radio in Panama," 13 March 1934, 819.76 Tropical Radio Company/69, DSNA 1930–39; Harold Hinton, "Welles: Our Man of the Hour in Cuba—His Successful Mediation Effort Was Aided by a Varied Experience in Dealing with Latin American Affairs," *New York Times Magazine*, 20 August 1933, pp. 111, 123; James B. Reston, "Acting Secretary: Sumner Welles, Career Diplomat Who Looks the Part Is Often Called to the White House These Days," *New York Times Magazine*, 3 August 1941, pp. 133, 143.

49. Merrell to Wilson, Memorandum, Division of Latin American Affairs, 20 April 1934, 819.74/283, and Merrell, "Interdepartmental Conference on Radio in Panama—July 2, 1934," Memorandum, Division of Latin American Affairs, 5 July 1934, 819.74/287 6/8, DSNA 1930–39.

50. "Our Isolation," editorial from *Star and Herald*, 26 June 1925, enclosed with South to Kellogg, 26 June 1925, 819.74/134, DSNA 1910–29. Davis to Stimson, "Radiotelegraph communications in the Republic of Panama," 21 October 1932, 819.74/23; American Minister Antonio Gonzalez to Hull, 25 August 1934, 819.76/18; Gonzalez to Hull, 26 September 1934, 819.76/20; Gonzalez to Hull, "Radio Broadcasting from Panama," 30 November 1934, 819.76/29; Assistant Secretary of State Sumner Welles to Secretary of War George H. Dern, 12 April 1935, 819.76/35; DSNA 1930–39.

51. Dawson to Hull, "Transmitting Publication of Panama Canal Containing 1936 Treaty and related Documents," 18 August 1939, 711.1928/851, DSNA 1930–39. See also Lester Langley, "Negotiating New Treaties with Panama: 1936," *Hispanic American Historical Review* 48.2 (1968), 229–31; Leonard, *Historical Dictionary of Panama*, 23–24.

52. American Minister George Summerlin to Hull, "Press comment upon treaty negotiations between the United States and Panama," 29 February 1936, 711.1928/521; Summerlin to Hull, "Opposition editorial objecting to treaty ratification in existing political situation," 6 March 1936, 711.1928/522; Summerlin to Hull, "United States Panama Treaty," 6 March 1936, 711.1928/523; Summerlin to Hull, "Further comment upon treaty between the United States and Panama" (includes clipping of the quoted *New York Herald Tribune* editorial), 10 March 1936, 711.1928/527; Sumerlin to Hull, "New treaty between the United States and Panama; statements attributed to Dr. Narciso Garay, one of the negotiators," 11 March 1936, 711.1928/528; Dawson to Hull, "Transmitting Publication of Panama Canal Containing 1936 Treaty and related Documents" (notes the ratification status of each treaty), 18 August 1939, 711.1928/851; DSNA 1930–39. See also Leonard, *Historical Dictionary of Panama*, 25, and Lester Langley, "The World Crisis and the Good Neighbor Policy in Panama, 1936–41," *The Americas* 24.2 (1967), 140.

53. The *New York Times* editorial is enclosed with Summerlin to Hull, "Further comment upon treaty between the United States and Panama," 10 March 1936, 711.1928/527; "A Policy that Gets Sweeter and Sweeter," editorial, *Green Bay Press-Gazette* (Green Bay, Wisconsin), 10 March 1936, p. 6. See also LaFeber, *The Panama Canal*, 67.

54. James T. Williams Jr., "Defense Peril Is Seen in Gagging Army," *Pittsburgh Sun-Telegraph*, 1 April 1936, p. 34. "Army and Navy to Guard Canal Despite Pact," *The Gazette* (Cedar Rapids, Iowa), 28 April 1936. "Army, Navy Give Notice to State Dept.," *The Times* (Munster, Indiana), 28 April 1928. "Navy to Guard Panama Canal No Matter What Diplomats Do," *Journal Times* (Racine, Wisconsin), 28 April 1936, p. 8. "Army, Navy Hold Safety of Canal Ahead of Treaty," *The Courier* (Waterloo, Iowa), 28 April 1936, p. 18. "Panama Pact Seen as Peril in War Time," *San Francisco Chronicle*, 28 April 1936, p. 2. Brigadier General Henry J. Reilly, "Panama Treaty Menaces Canal Safety, Says General Reilly," *Pittsburgh Sunday Telegraph*, 10 May 1936, p. 6. Rear Admiral Yates Stirling Jr., "New Treaty Puts U. S. Canal Defenses at Mercy of Panama, Warns Stirling," *San Francisco Examiner*, 17 May 1936, p. 30. Both the Reilly and Stirling essays were syndicated nationwide. Stirling's retirement is reported in "Nolan and Stirling

Dined; 'Thanks for Listening' Says President's Mother," *Brooklyn Times Union*, 16 April 1936, p. 17.

55. Senator Key Pitman, Chair of the Senate Foreign Relations Committee, to Hull, 22 April 1936, 711.1928/561 1/2, and White to Pittman, 3 August 1939, enclosed with Pittman to Hull, 10 August 1939, 819.74/352, DSNA 1930–39; "Daily Washington Merry Go Round," *Daily Press* (Newport News, Virginia), 27 March 1936, p. 4; "Canal Treaty with Panama Sidetracked: Navy and Senator Johnson Prevent Committee Action on State Department's Bill," *Hartford Courant* (Hartford, Connecticut), 3 November 1937, p. 5 (contains Johnson's "outrage" quote). "Panama Pact Menaced by Bipartisan Group of Senate Opponents," *Cincinnati Enquirer*, 12 February 1939, p. 1; William E. O'Connor of the Telecommunications Division to Chief of the Telecommunications Division Harvey Otterman, Memorandum, 3 September 1947, 819.74–347, DSNA 1945–49 (this 1947 memo recounts the overall history of US radio policy and notes White's critical role in killing the radio treaty); Langley, "The World Crisis," 141; Major, *Prized Possession*, 291–92; Howard A. Dewitt, "Hiram Johnson and Early New Deal Diplomacy, 1933–34," *California Historical Quarterly* 53.4 (1974), 377–86; "Hiram Johnson," *Oxford Reference*, https://www.oxfordreference.com/view/10.1093/oi/authority.20110810105158910. Johnson's vote totals in his 1934 reelection is from the "Our Campaigns" website at https://www.ourcampaigns.com/RaceDetail.html?RaceID=24097; Donald G. Godfrey and Louise M. Benjamin, "Radio Legislation's Quiet Backstage Negotiator: Wallace H. White, Jr.," *Journal of Radio Studies* 10.1 (2003), esp. 93; LaFeber, *Panama Canal*, 69–70.

56. Langley, "The World Crisis," 141; Sheila Hamilton, "Panamanian Politics and Panama's Relationship with the United States Leading up to the Hull Alfaro Treaty" (MA thesis, University of Victoria, 2009), 76; "Treaty Bogs Down in Listless Senate," *New York Times*, 25 July 1939, p. 6 (contains Johnson quote); Major, *Prized Possession*, 292; Pittman to Hull, 10 August 1939, 819.74/352, DSNA 1930–39.

57. Aubley B. Cowell, "Radio Markets in Panama," Special Report No. 12, pp. 1, 7, 16 August 1939, Folder "Foreign Service—Copies of Reports—Panama—1939—May—August," Box 388, Records Relating to Commercial Attaches' Reports, and Executive Vice President of Broadcasting Abroad Henry G. Hoberg to Chief of the Department of Commerce's Electrical Division John Payne, 31 January 1940, File "Radio—Latin America—1940–1943," Box 2482, Bureau of Foreign and Domestic Commerce General Records, 1914–58, Record Group 151, Records of the Bureau of Foreign and Domestic Commerce, United States National Archives, College Park, Maryland. Letter to Seldon Chapin, a lawyer for the State Department's Division of American Republics (formerly the Division of Latin American Affairs), 26 July 1939, 711.1928/819–1/2, DSNA 1930–39 (author's name illegible; contains "fine friendship" quote).

58. Major, *Prized Possession*, 259, 292; Langley, "The World Crisis," 147–52. Arosemena quoted in "Assures U. S. on New Pact: Arosemena Insists Our Troops Would Obtain Entry to Guard Canal," *Lexington Leader* (Lexington, Kentucky), 2 August 1937, p. 3. "The Menace of the New Treaty," *La Estrella de Panama* (Panama City), 30 April

1936, enclosed Summerlin to Hull, 1 May 1936, 711.1928/566, DSNA 1930–39. "Canal Treaty with Panama Sidetracked: Navy and Senator Johnson Prevent Committee Action on State Department's Bill," *Hartford Courant* (Hartford, Connecticut), 3 November 1937, p. 5. See also LaFeber, *The Panama Canal*, 70.

59. William E. O'Connor of the Telecommunications Division to Chief of the Telecommunications Division Harvey Otterman, Memorandum, 3 September 1947, 819.74–347, DSNA 1945–49.

60. Panama's pivot to the multilateral over bilateral agreements is mentioned in n. 14 that is appended to William E. O'Connor of the Telecommunications Division to Chief of the Telecommunications Division Harvey Otterman, Memorandum, 3 September 1947, in the version of the document available online at https://history.state.gov/historicaldocuments/frus1947v08/d814.

61. The Navy's prominent and then diminishing role in the development and control of American radio is a point of emphasis in Aitken, *The Continuous Wave*.

62. Hector H. Staff, *Historia y testimonios de la radiodifusion en Panama* (Universidad de Panamá, Facultad de Comunicación Social, Escuela de Radiodifusión, 1985), 32. *La Estrella de Panama* 1988 editorial quoted in Hector H. Staff, *La voz del barú y otros aspectos de radiodifusión y soberanía* (Panama: Mes de la Patria, 1988), 18–19. The website quotes are from http://emisorasdepanama.com/blog/historia-radio-panama and http://www.alonso-roy.com/era/era-46.html, respectively; see also https://panama-tour.site123.me/anécdotas-de-panamá/historia-de-la-radio-en-panamÁ, https://licas-comunicacionesinteractivas.blogspot.com/2009/05/httpwww.html, https://www.elistmopty.com/2020/09/historia-de-la-radio-en-panama.html, and http://historiadelaradioenpanama.blogspot.com/p/panama.html.

63. LaFeber, *The Panama Canal*, esp. ix, 78–81, 90, 98–102, 105–8, 111–12, 116, 136, 148, 150, 156, 215, 222.

Chapter 6. "An Almost Unbelievable Disregard of the Interests of the United States Listeners and Broadcasters"

1. "Cuba-U. S. Agreement Near: Compromise Is Expected to Avert Channel War," *Broadcasting*, 25 February 1946, 15, 78. "KFI Seeks to Keep Cuba Off 640kc," *Broadcasting*, 11 March 1946, 79. Stewart to Secretary of State James F. Byrnes, 24 February 1946, 837.76/2–2446, DSNA 1945–49.

2. On the uneasy coexistence of anti-Americanism with admiration of technology modernization in Cuba, see Louis A. Perez Jr., *On Becoming Cuban: Identity, Nationality, and Culture* (Chapel Hill: University of North Carolina Press, 2001), esp. 166–218.

3. Tehri Rantanen, *The Media and Globalization* (Thousand Oaks, CA: SAGE, 2005), 97. Guntram H. Herb, "National Identity and Territory," Guntram H. Herb and David H. Kaplan, eds., *Nested Identities: Nationalism, Territory and Scale* (Lanham, MD: Rowman and Littlefield 1999), 9; Liah Greenfeld, "The Origins and Nature of American Nationalism in Comparative Perspective," Knud Krakau, ed., *The American Nation—National Identity—Nationalism* (New Brunswick, NJ: LIT Vertag, 1997), 20, 47; Philip Resnick,

The Labyrinth of North American Identities (Toronto: University of Toronto Press, 2012), 102–3; Stephanie Kermes, *Creating an American Identity: New England, 1789–1825* (New York: Palgrave Macmillan, 2008), 1–2, 197; Colin Woodward, *American Nations: A History of the Eleven Rival Regional Cultures of North America* (New York: Penguin Books, 2011), 2.

4. Susan Douglas, "The Turn Within: The Irony of Technology in a Globalized World," *American Quarterly* 58.3 (2006), 619–38, esp. 620–21, 625–26, 637 (quoted portion from 637). Joseph A. Fry, "Place Matters: Domestic Regionalism and the Formation of American Foreign Policy," *Diplomatic History* 36.3 (2012), 452.

5. Christopher Sterling, "Clear Channel Stations: Powerful Major Market Radio Stations," 342–44, and Jim Hilliker, "KFI," 815–17, Christopher Sterling, ed., *The Encyclopedia of Radio* (New York: Fitzroy-Dearborn, 2004); "Clear Channel Decision Seen by Fall," *Broadcasting*, 22 April 1946, p. 90.

6. For the original NARBA terms, see Foust, *Big Voices of the Air*, 64–65; the treaty is also reprinted in *Broadcasting Yearbook*, 1941, 398–407. See also United States, *Limit Power of Radio Stations: Hearings Before the Committee on Interstate and Foreign Commerce*, 96.

7. The total number of standard broadcast stations is derived from lists of stations and broadcasters published in the 1935, 1941, and 1945 volumes of the *Broadcasting Yearbook*: for 1935, see p. 86; for 1941, see pp. 420–21; for 1945, see pp. 422–24. Nicholas Mendoza, Cuba's director of radio and NARBA delegation member, referenced the smaller figure of seventy-eight total stations in "Discussion in Habana regarding Cuban NARBA proposals," Assistant Chief of the State Department's Telecommunications Division Harvey Otterman to Chief of the State Department's Telecommunications Division France Colt de Wolf, 18 January 1946, 576.Washington/1–1846, DSNA 1945–49. For Cuba's position going into the conference, see also Meeting Minutes, Initiatives Committee, North American Regional Broadcasting Engineering Conference, 7 February 1946, 576.Washington/2–446 [49], DSNA 1945–49.

8. Perez, *On Becoming Cuban*, 25–28, 94–95, 166–218. Louis A. Perez Jr., *Cuba Under the Platt Amendment, 1902–1934* (Pittsburgh: University of Pittsburgh Press, 1986), xvi; Jules R. Benjamin, *The United States and the Origins of the Cuban Revolution: An Empire of Liberty in an Age of National Liberation* (Princeton, NJ: Princeton University Press, 1990), 63–65, 92–97; Yeidy M. Rivero, *Broadcasting Modernity: Cuban Commercial Television, 1950–1960* (Durham, NC: Duke University Press, 2015), 28, 32.

9. Rivero, *Broadcasting Modernity*, 23–44, esp. 31–33 (quoted portions of the 1939 decree from p. 31). "Discussion in Habana regarding Cuban NARBA proposals," Otterman to de Wolf, 18 January 1946, 576.Washington/1–1846, and Meeting Minutes, Initiatives Committee, North American Regional Broadcasting Engineering Conference, 7 February 1946, 576.Washington/2–446, DSNA 1945–49. *Broadcasting Yearbook* 1935, 86; *Broadcasting Yearbook* 1941, 420–21; *Broadcasting Yearbook* 1946, 536.

10. Perez, *Cuba Under the Platt Amendment*, 339; Walter LaFeber, *The American Age: U.S. Foreign Policy at Home and Abroad, 1750 to the Present*, 2nd ed. (New York: W. W. Norton, 1996), 377–78; Peter Smith, *Talons of the Eagle: Dynamics of U.S.-Latin American Relations*, 2nd ed. (New York: Oxford University Press, 1999), 70; Thomas G. Paterson, *Contesting*

Castro: The United States and the Triumph of the Cuban Revolution (New York: Oxford University Press, 1994), 61; Benjamin, *The United States and the Origins of the Cuban Revolution*, 96–105 (US Ambassador George Messersmith quoted in Benjamin on p. 98); Mark T. Gilderhus, *The Second Century: U.S.-Latin American Relations since 1889* (Wilmington, DE: Scholarly Resources, 2000), 97–98.

11. Proposals Submitted by the Delegation of Cuba (Documents 6 and 11), 576. Washington/2–446, DSNA 1945–49; "Eight Nations Attend NARBA Meeting," *Broadcasting*, 4 February 1946, p. 18; "Cuba Protesting Present NARBA Rule," *Radio Daily*, 8 February 1946, pp. 1, 4; "Cuba Firm, NARBA Enters Third Week," *Broadcasting*, 18 February 1946, p. 17.

12. Cuba's staunch opposition to the 650-mile rule is reflected in the considerable discussion devoted to it during the meetings of the Technical Committee; for example, see the Technical Committee's meeting minutes from 5 February 1946 (File 89338), 7 February 1946 (File 89403), and 8 February 1946 (File 89413), 576.Washington/2–446; Cuba's insistence on securing rights to broadcast over 640kc is addressed in the Technical Committee's meeting minutes from 13 February 1946 (File 89667), 576.Washington/2–446; DSNA 1945–49. See also "Cuba Protesting Present NARBA Rule," *Radio Daily*, 8 February 1946, pp. 1, 4; "Cuba-U. S. Agreement Near: Compromise Is Expected to Avert Channel War," *Broadcasting*, 25 February 1946, pp. 15, 78.

13. Meeting Minutes, Technical Committee, North American Regional Broadcasting Conference, esp. pp. 4–5, 15 February 1946 (File 89304), 576.Washington/2–446, DSNA 1945–49; "Cuba Firm, NARBA Enters Third Week," *Broadcasting*, 18 February 1946, 17; "Cuba's NARBA Victory Portends U. S. Row," *Broadcasting*, 4 March 1946, p. 80; Raymond Guy quoted in *Limit Power of Radio Stations: Hearings Before the Committee on Interstate and Foreign Commerce*, 592; E. K. Jett to Secretary of State James F. Byrnes, 16 May 1946, reprinted in United States, *Foreign Relations of the United States, 1946: Volume 11, The American Republics* (Washington, DC: US Government Printing Office, 1969), 737–39 (Jett quoted on p. 738).

14. Meeting Minutes, Technical Committee, North American Regional Broadcasting Conference, esp. pp. 10–11, 15 February 1946 (File 89304), 576.Washington/2–446, DSNA 1945–49; "Cuba Firm, NARBA Enters Third Week," *Broadcasting*, 18 February 1946, p. 17; "Cuba-U. S. Agreement Near at NARBA," *Broadcasting*, 25 February 1946, pp. 15, 78; "Cuba's NARBA Victory Portends U. S. Row," *Broadcasting*, 4 March 1946, pp. 17, 80; for Cuba's "grapevine hints," see Louis Caldwell (Clear Channel Broadcasting Service lawyer and lobbyist) to Edwin Craig (owner and manager of Nashville clear channel station WSM), 22 March 1946, reprinted in *Limit Power of Radio Stations: Hearings Before the Committee on Interstate and Foreign Commerce*, 61.

15. De Wolf quoted in "U. S. Seeks Continuance of NARBA for Two Years," *Radio Daily*, 7 February 1946, 7. For Jett's position, see "Cuba Protesting Present NARBA Rule," *Radio Daily*, 8 February 1946, p. 4; "NARBC to Conclude Today; Cuba Threatens to Resign," *Radio Daily*, 25 February 1946, 1, 7. "Cuba Maneuver," *Broadcasting*, 18 February 1946, p. 54. On *Broadcasting*'s editorial slant, see McChesney, *Telecommunications, Mass Media, and Democracy*, 110–11.

16. "Cuba's NARBA Victory Portends U. S. Row," *Broadcasting*, 4 March 1946, pp. 17, 80; the distinctions between the original treaty and the interim agreement are addressed in *Limit Power of Radio Stations: Hearings Before the Committee on Interstate and Foreign Commerce*, 60, 613, 615, and 657.

17. Caldwell articulated his views at the *Limit Power of Radio Stations: Hearings Before the Committee on Interstate and Foreign Commerce*, 58, 65–66, 100. For Craig, see Commercial Attaché in Cuba to Secretary of State James F. Byrnes, "Cuban Press Comments Concerning the North American Regional Broadcasting Conference," 8 March 1946, 576.Washington/3–846, DSNA 1945–49, and "NARBA Views of Clears, Regionals," *Broadcasting*, 4 March 1946, pp. 17, 78.

18. George C. Howard (the American commercial attaché in Havana) to Byrnes, 8 March 1946, 576.Washington/3–846, DSNA 1945–49. For the background on *Alerta*, see K. Lynn Stoner, "Militant Heroines and Consecration of the Patriarchal State," *Cuban Studies* 34 (2003), 84; Carlos Todd, "The Communist Destruction of the Free Press in Cuba," *Communist Penetration and Exploitation of the Free Press*, study prepared for the Subcommittee to Investigate the Administration of the Internal Security Act and Other Internal Security Laws of the Committee on the Judiciary, United States Senate, Eighty-Seventh Congress, Second Session (Washington, DC: US Government Printing Office, 1962), 28.

19. Frances Colt de Wolf, memorandum summarizing the NARBA conference (includes attachment of de Wolf's closing remarks to the conference), 1 March 1946, 576. Washington/3–146, DSNA 1945–49; "Cuba's NARBA Victory Portends U. S. Row," *Broadcasting*, 4 March 1946, pp. 17, 80; Jett to Byrnes, 16 May 1946 (see n. 13 above).

20. Foust, *Big Voices of the Air*, 201. Ryan quoted in "Cuba's NARBA Victory Portends U. S. Row," *Broadcasting*, 4 March 1946, p. 80. Interference from Cuba addressed in "KFI Seeks to Keep Cuba Off 640kc," *Broadcasting*, 11 March 1946, p. 79. Eugene Jervis to the Federal Communications Commission, 7 March 1946, enclosed with Eugene Jervis to Francis Colt de Wolf of the Department of State's Telecommunications Division, 6 March 1946, 837.76/3–646, DSNA 1945–49.

21. The bulk of the correspondence referenced above, discussed in greater detail below, can be found in the DSNA 1945–49 files spanning the decimal number 837.76/2–2446 through 837.76/2–2546, including Goodall to Byrnes, 23 February 1946, 837.76/2–2446. See also FCC Commissioner Roy Wakefield to John W. Costello of the Los Angeles Chamber of Commerce, 19 April 1946, File 194–61 (Clear Channel Station Hearing, 5–9–45, Docket 6741, Protests and Endorsements), Box 493, FCC Records 1927–46. The stream of protests from California are also acknowledged in "Cuba's NARBA Victory Portends U. S. Row," *Broadcasting*, 4 March 1946, p. 80. Quotes from the California Farm Bureau are from "Resolution Passed by California Farm Bureau Federation in San Francisco, California, 1946," reprinted in *Limit Power of Radio Stations: Hearings Before the Committee on Interstate and Foreign Commerce*, 223.

22. On the economy and unemployment in the West, see Earl Pomeroy, *The Pacific Slope: A History of California, Oregon, Washington, Idaho, Utah, and Nevada* (Lincoln: Univer-

sity of Nebraska Press, 1991), 301, 313. On California agriculture's significance to the nation, see "Statement of the California Farm Bureau Federation on S. 2232 Presented to the Senate Committee on Interstate and Foreign Commerce by Vice President of the California Farm Bureau Federation Clyde O. Hoober," 7 April 1948, *Limit Power of Radio Stations: Hearings Before the Committee on Interstate and Foreign Commerce*, 301–2; Richard W. Etulain and Michael P. Malone, *The American West: A Modern History, 1900 to the Present*, 2nd ed. (Lincoln: University of Nebraska Press, 2007), 22–23; and "California Agriculture: Feeding the Future," issued by the Governor's Office of Planning and Research California Rural Policy Task Force (2003), http://rural.legislature.ca.gov/sites/rural.legislature.ca.gov/files/OPR_report.pdf. On the specific significance of Los Angeles and surrounding counties, see Abraham F. Lowenthal, *Global California: Rising to the Cosmopolitan Challenge* (Stanford, CA: Stanford University Press, 2009), esp. 25, 31. For the specific ranking of counties, see United States, *United States Census of Agriculture, 1945* (Washington, DC: US Government Printing Office, 1947), Table 2, "Value of Farm Products Sold or Used by Farm Households—100 Leading Counties, 1944, with Comparisons, 1939," p. 2.

23. United Date Growers Association (Indio, California) to Byrnes, 23 February 1946, and Badger to Byrnes, 24 February 1946, 837.76/2–2446, DSNA 1945–49. "Clear Channel Decision Seen by Fall," *Broadcasting*, 22 April 1946, 18, 90–91. On the Depression-era politics and antiunion efforts, see Gary Nash, *The American West Transformed: The Impact of the Second World War* (Bloomington: Indiana University Press, 1985), 46–47; Cletus E. Daniel, *Bitter Harvest: A History of California Farmworkers, 1870–1941* (Ithaca, NY: Cornell University Press, 1981), esp. 251–52, 270–71, 282; Foust, *Big Voices of the Air*, 191, 196–202.

24. The phrasing choices in this sentence reflect the influence of Brooke L. Blower, "From Isolationism to Neutrality: A New American Framework for Understanding American Political Culture, 1919–1941," *Diplomatic History* 38.2 (2014), 345–76. Blower is rightly critical that the standard juxtaposition of isolationism versus internationalism to describe the prewar political dispositions in various parts of the country obscures the diversity of views among those frequently lumped together under those respective umbrellas. For that reason, this chapter casts the Far West not with the typical "isolationist" label but as a region whose population favored neutrality and nonintervention; likewise, the qualified reference to the "*purported* internationalism and more interventionist dispositions associated with the urban areas of the country and the federal government" is also intended as an implicit acknowledgment that such standard blanket descriptions of popular attitudes obscure the complexity of views surrounding this fraught period of US foreign relations.

25. On the Far Western identity, see Woodard, *American Nations*, 12, 243–53 (esp. 244, 251–53). For the industrial transformation of the United States, rural concerns about loss of autonomy, animosity toward the expanding federal bureaucracy, and the growing manifestation of a urban-rural divide, see Don S. Kirschner, *City and Country: Rural Responses to Urbanization in the 1920s* (Westport, CT: Greenwood Press, 1970), esp. 23–56; Ronald R. Kline, *Consumers in the Country: Technology and Social Change in Rural America* (Baltimore: Johns Hopkins University Press, 2000), 13; James H. Shideler,

"'Flappers and Philosophers,' and Farmers: Rural-Urban Tensions of the Twenties," *Agricultural History* 47.4 (1973), 283–99; David B. Danbom, *Born in the Country: A History of Rural America*, 2nd ed. (Baltimore: Johns Hopkins University Press, 2006), 158, 170–75, 221. For the rural Far West's isolationism, see Pomeroy, *The Pacific Slope*, 291; Edward W. Chester, *Sectionalism, Politics, and American Diplomacy* (Metuchen, NJ: Scarecrow Press, 1975), 235.

26. Alfred E. Eckes Jr., *Opening America's Market: U. S. Foreign Trade Policy Since 1776* (Chapel Hill: University of North Carolina Press, 1995), 255–56.

27. George J. Richardson Breeding Farm and Hatchery Company (San Gabriel, California) to Byrnes, 23 February 1946; Herdman to Byrnes, 24 February 1946; United Date Growers Association to Byrnes, 23 February 1946; and Isham to Byrnes, 23 February 1946, 837.76/2–2446, DSNA 1945–49.

28. Arnold Offner, "'Another Such Victory': President Truman, American Foreign Policy, and the Cold War," *Diplomatic History* 23.2 (1999), esp. 131, 137. For intersection between isolationism and unilateralism, see Christopher McKnight Nichols, *Promise and Peril: America at the Dawn of a Global Age* (Cambridge, MA: Harvard University Press, 2011), 348–50.

29. Isham to Byrnes, 23 February 1946; Ed Brunton (Mentone, California) to Byrnes, 24 February 1946; Ford to Byrnes, 24 February 1946; and James Leonard (Redlands, California) to Byrnes, 24 February 1946, 837.76/2–2446, DSNA 1945–49.

30. Linda Noel, *Debating American Identity: Southwest Statehood and Mexican Immigration* (Tucson: University of Arizona Press, 2014), 17–18, 53–55, 61, 84, 99, 106–7, 122–25, 136, 142; Mae Ngai, *Impossible Subjects: Illegal Aliens and the Making of Modern America*, rev. ed. (Berkeley: University of California Press, 2014), 64–75.

31. Gerd Baumann, *The Multicultural Riddle: Rethinking National, Ethnic, and Religious Identities* (New York: Routledge, 1999), 21; Peter J. Burke, "Relationships Among Multiple Identities," Peter J. Burke, Timothy J. Owens, Richard T. Sherpe, and Peggy Thoits, eds., *Advances in Identity Theory and Research* (New York: Kluwer Academic/Plenum Publishers, 2003), esp. 201–2; Simon Coleman and Peter Collins, "Introduction: Ambiguous Attachments: Religion, Identity, and Nation," Simon Coleman and Peter Collins, eds., *Religion, Identity, and Change: Perspectives on Global Transformations* (Burlington, VT: Ashgate, 2004), 4–6; Sheldon Stryker and Peter J. Burke, "The Past Present and Future of an Identity Theory," *Social Psychology Quarterly* 63.4 (2000), esp. 287, 289, 292; Peggy A. Thoits, "Personal Agency in Multiple Role Identities," *Advances in Identity Theory and Research*, esp. 181–83.

32. B. J. Palmer (WHO in Des Moines, Iowa) to Byrnes, 25 February 1946, 837.76/2–2546; Glenn Snyder (WLS in Chicago, Illinois) to Byrnes, 25 February 1946, 837.76/2–2346; President of the Illinois Agricultural Association Charles Schuman (arguing on behalf of station WLS) to Byrnes, 24 February 1946, 837.76/2–2446; Craig (WSM in Nashville Tennessee) to Byrnes, 25 February 1946, 837.76/2–2546; Hugh Halff (WOAI in San Antonio, Texas) to Byrnes, 25 February 1946, 837.76/2–2546; J. D. Shouse (WLW in Atlanta, Georgia) to Byrnes, 26 February 1946, 837.76/2–2356; DSNA 1945–49.

33. Foust, *Big Voices of the Air*, 197–202; *Limit Power of Radio Stations: Hearings Before*

the Committee on Interstate and Foreign Commerce documents several statements of Farm Bureau and affiliated groups' support dating back to the early 1940s: American Farm Bureau statement in favor of Clear Channel Broadcasting, December 1946, pp. 303–4; statement of Director of the Texas Farm Bureau Loys D. Barbour, 7 April 1948 (Barbour dates his organization's opposition to any dismantling of the clear channel system to 1943), pp. 305–7; Eastern Dark-Fired Tobacco Growers Association to WSM Broadcasting Company Manager Harry Stone, 29 March 1946, p. 377; Executive Secretary of the Tennessee Farm Bureau Federation O. B. Long to Harry Stone, 29 March 1946, p. 379; statement of Director of Information for the Kentucky Farm Bureau Joe Betts, 7 April 1948, pp. 383–86; Assistant Executive Secretary of the Michigan Farm Bureau J. F. Yaeger to Chairman of the Senate's Interstate and Foreign Commerce Committee Charles Tobey (RNH) (Yaeger references the Michigan Farm Bureau's support of the clear channel system dating back to a 1942 resolution passed by the organization), pp. 462–63; Georgia farm support, including from county-level Farm Bureaus, pp. 1213–16; President of the Illinois Agricultural Association Charles B. Shuman to Senator Tobey, 31 March 1948, p. 1243.

34. These groups' concerns about the clear channel system were registered during the 1948 hearings documented in *Limit Power of Radio Stations: Hearings Before the Committee on Interstate and Foreign Commerce.* See President of the Iowa Farm Bureau Federation E. Howard Hill to Chairman of the Senate's Interstate and Foreign Commerce Committee Charles Tobey (RNH), 9 April 1948, p. 928; "A Statement Made by Michigan State College and Radio Station WKAR Prepared for Presentation at the Clear Channel Hearing Before the Federal Communications Commission," 31 October 1947, pp. 1320–21; President of Michigan State College of Agriculture and Applied Science John Hannah to Senator Tobey, 12 April 1948, pp. 1314–15; Statement of Richard M. Fairbanks in Favor of the Johnson Bill in Behalf of the Indiana Broadcasting Company, 14 April 1948, pp. 938–46; President of the Indiana Farm Bureau Hassil E. Schenk to Senator Homer E. Capehart (RIN), 25 April 1948, pp. 1254–55; Secretary of the Ohio Farm Bureau Federation John W. Sims to Senator Tobey, 20 April 1948, p. 1105; President of the Ohio Farm Bureau Federation Perry Green to Senator Wallace White, 29 April 1948, p. 1442. For the WOIKFI conflict, see Foust, *Big Voices of the Air,* 192–95.

35. "Channel Compromises to Cuba Called Regrettable by de Wolf," *Broadcasting,* 4 March 1946, 75; "NARBA Views of Clears, Regionals," *Broadcasting,* 4 March 1946, 17, 76. On the position of different broadcasting groups toward the 1946 NARBA, see *Limit Power of Radio Stations: Hearings Before the Committee on Interstate and Foreign Commerce,* 63–64, 92.

36. For reports of interference, see Second Annual Report, North American Regional Broadcasting Engineering Committee, Fiscal Year 1948, p. 2 of the main document, pp. 2, 8 of appendix 1 (Radiation Measurements of Stations CMQ, CMCY, and CMHQ by the North American Regional Broadcasting Engineering Committee), and pp. 3, 10 of appendix 2 (Horizontally Polarized Radiation from Closed Space Arrays),

file "Mr. Joseph Kittner," Office of the Chief Engineer, Records Relating to the Negotiation of International Agreements, Treaties, and Conferences, 1875–1967, Box 26, FCC Records 1927–46. The assertion that intermittent interference that followed the 1946 NARBA revision never provoked complaints of impending radio doom like those predictions that preceded the revision is based on the fact that no newspapers, from the *Los Angeles Times* to any of the local papers within KFI's range, appear to have published any stories or reports about Cuba causing interference over 640kc following the interim agreement's enactment.

37. Statement by Isham, 7 April 1948, and Isham to Chairman of the FCC Wayne Coy, 10 March 1948, reprinted in *Limit Power of Radio Stations: Hearings Before the Committee on Interstate and Foreign Commerce* 221, 231, 283. Testimony and support from KFI advocates appear on pp. 207–303. The nearly one hundred pages that this documentation comprises far exceeds the amount of space and time used by any other group testifying for or against this bill. Statement of Frank P. Schreiber (station WGN of Chicago, Illinois), 8 April 1948, *Limit Power of Radio Stations: Hearings Before the Committee on Interstate and Foreign Commerce*, 437. For the perceived threat posed by the 1950 NARBA revision, see "U. S. and Cuba in Radio War," *Times-Herald* (Olean, New York), 17 December 1949, p. 1. "Manguson Opposed to Cuban Requests," *Medford Mail Tribune* (Medford, Oregon), 8 December 1949, p. 6; "GOP Aroused over Cuban Radio Pact," *Bakersfield Californian* (Bakersfield, California), 8 December 1949, p. 30. For more on the 1948 Senate hearings, see Foust, *Big Voices of the Air*, 201–2. For opposition and delays in approving the 1950 NARBA revision, see Christopher Sterling, "North American Regional Broadcasting Agreement: Sharing Frequencies Among the United States, Canada, Mexico, and the Caribbean," Sterling and O'Dell, eds., *The Concise Encyclopedia of American Radio*, 528.

38. Benjamin, *The United States and the Origins of the Cuban Revolution*, 121–87; Timothy B. Tyson, *Radio Free Dixie: Robert F. Williams and the Roots of Black Power* (Chapel Hill: University of North Carolina Press, 1999), 287–94; Daniel Walsh, *An Air War with Cuba: The United States Radio Campaign Against Castro* (Jefferson, NC: McFarland and Company, 2012), 16–27, 62–72, 88–102, 206–14.

39. Fry, "Place Matters," 452.

Conclusion

1. Sweet to Secretary of State Charles Evans Hughes, 27 March 1923, 812.74/189, DSNA 1910–29.

2. James Wood, *History of International Broadcasting*, vol. 2 (London: Institution of Electrical Engineers, 2000), 19; Alan L. Heil Jr., *Voice of America: A History* (New York: Columbia University Press, 2003), 4, 69–71.

3. John Sinclair, "Culture and Trade: Some Theoretical and Practical Considerations," Emile G. McAnany and Kenton T. Wilkinson, eds., *Mass Media and Free Trade: NAFTA and the Cultural Industries* (Austin: University of Texas Press, 1996), 42–43; Giuseppe Richeri, "Global Film Market, Regional Problems," *Global Media and China* 1.4 (2016), 313; Hernan Galperin, "Cultural Industries Policy in Regional Trade Agree-

ments: The Cases of NAFTA, the European Union and MERCOSUR," *Media, Culture, and Society* 21.5 (1999), 627; and Drew Alexander Ross, "The Impact of the International Box Office on Movie Making Decisions," *Hollywood Insider*, 15 August 2020, https://www.hollywoodinsider.com/international-box-office-impact/.

4. For communications satellites and their integration into global communications infrastructures, see Slotten, *Beyond Sputnik and the Space Race*. Slotten also notes the increasing prominence and effectiveness of undersea fiberoptic cables on pp. 187–88. Johnson's quote is from Schwoch, *Global TV*, 152.

5. Andrew J. Nathan, *Chinese Democracy* (Berkeley: University of California Press, 1986), 152–92.

6. The *Jurassic Park* information is from (the chapter title notwithstanding) Nestor García Canclini, "North Americans or Latin Americans: The Redefinition of Mexican Identity and the Free Trade Agreements," McAnany and Wilkinson, eds., *Mass Media and Free Trade*, 150.

7. Leah Greismann, "After the Surprise Success of 'Like Water for Chocolate,'" 30 October 1995, UPI Archives, https://www.upi.com/Archives/1995/10/30/After-the-surprise-success-of-Like-Water-for-Chocolate/3191815029200/. The redubbing of English-language scenes with different spoken English is noted on the film's *IMDB* ("Internet Movie Database") page, https://www.imdb.com/title/tt0103994/alternateversions?tab=cz&ref_=tt_trv_alt). Shirley Li, "How Hollywood Sold Out to China," *The Atlantic*, 10 September 2021, https://www.theatlantic.com/culture/archive/2021/09/how-hollywood-sold-out-to-china/620021. Nick Hemming, "Movies Hollywood Was Forced to Change for the Middle East," *Looper*, 11 December 2020, https://www.looper.com/188150/movies-hollywood-was-forced-to-change-for-the-middle-east.

8. Gilbert Gagné, "The Treatment of Cultural Products and the Cultural Exemption Clause," Gilbert Gagné and Michèle Rioux, eds., *NAFTA 2.0: From the First NAFTA to the United-States-Mexico-Canada Agreement* (New York: Palgrave MacMillan, 2022), esp. 244, 247, 251–52, 254, 256 (cultural exemption clause wording from p. 254, emphasis added); Emile G. McAnany and Kenton T. Wilkinson, "Introduction," McAnany and Wilkinson, eds., *Mass Media and Free Trade*, 10–11; Galperin, "Cultural Industries Policy in Regional Trade Agreements," 628, 631–34; Roger de la Garde, "There Goes the Neighborhood: Montréal's Free Television Market and Free Trade," McAnany and Wilkinson, eds., *Mass Media and Free Trade*, 272–74. Gilbert Gagné, "Cultural Sovereignty, Identity, and North American Integration," *Quebec Studies* 36.1 (2003), 33–35, 42; Carolyn P. Egri, David A. Ralston, Cheryl S. Murray, and Joel D. Nicholson, "Managers in the NAFTA Countries: A Cross-Cultural Comparison of Attitudes Toward Upward Influence Strategies," *Journal of International Management* 6.2 (2000), 154.

9. Janna Anderson and Lee Rainee, "Digital Live in 2025—Summary: 15 Theses About the Digital Future," *Pew Research Center*, 11 March 2014, https://www.pewresearch.org/internet/2014/03/11/digital-life-in-2025.

10. Matteo Cinelli, Gianmarco De Francisci Morales, Alessandro Galeazzi, Walter Quattrociocchi, and Michele Starnini, "The Echo Chamber Effect on Social Media," *Proceedings of the National Academy of Sciences* 118.9 (2021), via https://www.pnas.org/

doi/abs/10.1073/pnas.2023301118; Michalis Mamakos and Eli J. Finkel, "The Social Media Discourse of Engaged Partisans Is Toxic Even When Politics Are Irrelevant," *PNAS Nexus* 2.10 (2023), via https://academic.oup.com/pnasnexus/article/2/10/pgad325/7293179.

11. "No Way, Jose" campaign ad, https://x.com/kayiveyforgov/status/1513604794929328134?s=20.

12. Leah Askarinam, "Kay Ivey Races to the Right in Alabama," *New York Times*, 26 April 2022, https://www.nytimes.com/2022/04/26/us/politics/kay-ivey-trump-alabama.html; Joe Skolnik, "'No way, José': Alabama Governor Sparks Cries of Racism with New Campaign Ad Attacking Joe Biden," *Salon*, 15 April 2022, https://www.salon.com/2022/04/15/no-way-jos-alabama-governor-sparks-cries-of-with-new-campaign-ad-attacking-joe-biden. Ivey defended herself on the April 18 news broadcast on WVTM, an Alabama-based NBC subsidiary: https://www.youtube.com/watch?v=v3n9hx-0AKI&t=7s. Franklin quoted in Baron, *The English-Only Question*, 66. Hector Amaya, "Nativist Liberalism and the Disciplining of Spanish Language Media," esp. 28.

Bibliography

Archival Sources

Acadian Handicraft Project, Special Collections, Hill Memorial Library, Louisiana State University, Baton Rouge, Louisiana.

John R. Brinkley Papers, Kansas State Historical Society, Topeka, Kansas.

Arthur Carson Papers, Kansas State Historical Society, Topeka, Kansas.

George H. Clark Radioana Collection, Archives Center at the National Museum of American History at the Smithsonian Institution.

General Education Board Appropriations Records, Rockefeller Foundation Archives, Sleepy Hollow, New York.

General Electric Papers, Schenectady Museum Archives, Schenectady, New York.

Herbert Hoover Commerce Department Papers, Herbert Hoover Presidential Library, West Branch, Iowa.

Louisiana State University Office of the Chancellor Records, Special Collections, Hill Memorial Library, Louisiana State University, Baton Rouge, Louisiana.

National Advisory Council on Radio in Education Records, New York Public Library, New York, New York.

United States Department of State Central Decimal Files, Record Group 59, National Archives at College Park, Maryland.

United States Bureau of Foreign and Domestic Commerce Records, Record Group 151, United States National Archives, College Park, Maryland.

United States Federal Communications Commission Records, Office of the Executive Director, General Correspondence, 1927–46, Record Group 173, United States National Archives, College Park, Maryland.

Rockefeller Foundation Records, Rockefeller Archives Center, Sleepy Hollow, New York.

Office of the Messers. Rockefeller Records, Rockefeller Archives Center, Sleepy Hollow, New York.

Published Primary Sources

Bennet, Ira. *History of the Panama Canal: Its Construction and Builders*. Washington, DC: Historical Publishing Company, 1915.

Blake, Katherine D. "The New Outlook." *Journal of Education* 96.3 (1922): 75.

Buell, Raymond Leslie. "Union or Disunion in Central America?" *Foreign Affairs* 11.3 (1933): 478–89.

"Eighth Institute for Education by Radio." *Education By Radio: A Bulletin to Promote the Use of Radio for Educational, Cultural, and Civic Purposes* 7.6 (1937): 23–25.

EMF Electrical Year Book: An Encyclopedia of Current Information about Each Branch of the Electrical Industry, with a Dictionary of Electrical Terms and a Classified Directory of Electrical and Related Products and Their Manufacturers in the United States and Canada. Chicago: Electrical Trade Publishing Company, 1923.

Engel, E.F. "The Broadcasting of Modern Foreign Languages in the United States: Second Survey." *Modern Language Journal* 22.8 (1938): 626–28.

Engel, E.F. "Radio as a Medium of Modern Language Instruction." *Modern Language Journal* 20.3 (1935): 165–71.

Fosdick, Raymond. *The Story of the Rockefeller Foundation*. New York: Harper and Row, 1952.

Howard, Charles A. "Electrification for Latin America." *Electrical Engineering* 63.2 (1944): 43–46.

Lefevre, Enrique. *Más allá del olvido: Ernesto Lefevre y el imperialismo yanky*. Panama City: Imprenta Universidad de Panama, 1972.

Lazarsfeld, Paul F. *Radio and the Printed Page: An Introduction to the Study of Radio and Its Role in the Communication of Ideas* (New York: Duell, Sloan, and Pierce, 1940).

Mahan, Alfred Thayer. *The Interest of America in Sea Power, Present and Future*. New York: Little, Brown, 1897.

Marx, Groucho. *Groucho and Me*. New York: B. Geis Associates, 1959.

Mawhinney, Clara Krefting and Harley A. Smith. *Business and Professional Speech*. New York: American Book Company, 1950.

Mercer, C.H. "Radio and Aural Comprehension." *Modern Language Journal* 15.5 (1931): 325–36.

Missouri. *Appendix to the House and Senate Journals of the General Assembly of State of Missouri.* Jefferson City, MO: Midland Printing Company, 1933.

Morgan, B.Q. and John Van Horne. "Bibliography of Modern Language Methodology for 1924." *Modern Language Journal* 9.8 (1925): 495–501.

Munro, Dana Gardner. "Book Review: *Career Ambassador* by William L. Beaulac." *Hispanic American Historical Review* 32.1 (1952): 116–17.

——. *A Student in Central America.* New Orleans: Tulane University Press, 1983.

Ogden, C.K. "Will Basic English Become the Second Language." *Public Opinion Quarterly* 8.1 (Spring 1944): 3–9.

Olivier, Louise. "A Glossary of Variants from Standard-French in St. Landry Parish." MA thesis, Louisiana State University, 1937.

"The Radio Panorama." *Education By Radio: A Bulletin to Promote the Use of Radio for Educational, Cultural, and Civic Purposes* 7.7 (1937): 29–34.

Richards, I.A. "Basic English and Its Applications." *Journal of the Royal Society of the Arts* 87.4515 (1939): 735–55.

——. *Basic English and Its Uses.* New York: W.W. Norton, 1943.

Richards, I.A. and Christine Gibson, "Learning Basic English: An Over-all View." *English Journal* 34.6 (1945): 303–9.

Smith, Harley Albert. "A Recording of English Sounds at Three Age Levels in Ville Platte, Louisiana." PhD dissertation: Louisiana State University, 1936.

Smith, Harley A. and Ida Lee King Wilbert, *Everyday English.* Austin, Texas: Steck-Vaughn Company, 1965.

Todd, Carlos. "The Communist Destruction of the Free Press in Cuba." *Communist Penetration and Exploitation of the Free Press* [study prepared for the Subcommittee to Investigate the Administration of the Internal Security Act and Other Internal Security Laws of the Committee on the Judiciary, United States Senate, Eighty-Seventh Congress, Second Session]. Washington, DC: US Government Printing Office, 1962.

Underwood, Eric. "The Radio University in Peace and War." *American Scholar* 14.1 (1944): 86–96.

United States. *Communist Penetration and Exploitation of the Free Press* [study prepared for the Subcommittee to Investigate the Administration of the Internal Security Act and Other Internal Security Laws of the Committee on the Judiciary, United States Senate, Eighty-Seventh Congress, Second Session]. Washington, DC: US Government Printing Office, 1962.

United States. *Foreign Relations of the United States, 1946: Volume 11, The American Republics.* Washington, DC: US Government Printing Office, 1969.

United States. *Limit Power of Radio Stations: Hearings Before the Committee on Interstate and Foreign Commerce United States Senate, Eightieth Congress, Second Session on S. 2231. A Bill to Limit AM Radio Broadcast Stations to 50,000 Watts and to Provide for Duplication of Clear Channels.* Washington, DC: US Government Printing Office, 1948.

United States. *United States Census of Agriculture, 1945.* Washington, DC: US Government Printing Office, 1947.

United States, Department of Commerce, Bureau of the Census. *Historical Statistics of the United States, Colonial Times to 1970*, part 2. Washington, DC: US Government Printing Office, 1975.

United States, Department of Commerce, Bureau of Navigation. *Amateur Radio Stations of the United States.* Washington, DC: US Government Printing Office, 1926.

United States, Library of Congress. *Background Documents Relating to the Panama Canal.* Washington, DC: US Government Printing Office, 1977.

United States, Library of Congress. *A List of American Doctoral Dissertations Printed in 1918*. Washington, DC: US Government Printing Office, 1921.

Wilson, Charles Morrow. *Empire in Green and Gold: The Story of the American Banana Trade*. New York: Holt, Reinhart, and Winston, 1947.

Secondary Sources

Abramsky, Sasha. "The Anglo-American Misalliance." *World Policy Journal* 25.1 (2008): 72–79.

Adas, Michael. *Dominance by Design: Technological Imperatives and America's Civilizing Mission*. Cambridge, MA: Belknap Press of Harvard University Press, 2006.

———. *Machines as the Measure of Men: Science, Technology, and Ideologies of Western Dominance.* Ithaca, NY: Cornell University Press, 1989.

Aitken, Hugh G.J. *The Continuous Wave: Technology and American Radio, 1900–1932*. Princeton, NJ: Princeton University Press, 1985.

Amaya, Hector. "Nativist Liberalism and the Disciplining of Spanish Language Media." *Communication and Power in the Global Era*, edited by Marwan M. Kraidy, 14–31. New York: Taylor and Francis, 2013.

Ancelet, Barry Jean, Jay D. Edwards, and Glen Pitre. *Cajun Country*. Jackson: University of Mississippi Press, 1991.

Anderson, Benedict. *Imagined Communities: Reflections on the Origins and Spread of Nationalism*, rev. ed. New York: Verso Books, 1991.

Anderson, Stuart. *Race and Rapprochement: Anglo-Saxonism and Anglo-American Relations, 1895–1904*. Rutherford, NJ: Fairleigh Dickinson University Press, 1981.

Arceneaux, Noah. "Acadian Airwaves: A History of Cajun Radio." *Journal of Radio and Audio Media* 30.2 (2023): 664–79.

Badwan, Khawla. *Language in a Globalised World: Social Justice Perspectives on Mobility and Contact*. New York: Palgrave Macmillan, 2021.

Baker, Jr., George W. "The Wilson Administration and Panama, 1913–1921." *Journal of Inter-American Studies* 8.2 (1966): 279–93.

Balderrama, Francisco E. and Raymond Rodriguez. *Decade of Betrayal: Mexican Repatriation in the 1930s*, rev. ed. Albuquerque: University of New Mexico Press, 2006.

Bankston, Carl L. and Jacques Henry, "Endogamy among Louisiana Cajuns: A Social Class Explanation." *Social Forces* 77.4 (1999): 1317–38.

Barnett, George A., Thomas Jacobson, Young Choi, and Sulien Sun-Miller, "An Examination of the International Telecommunication Network." *Journal of International Communication* 3.2 (1996): 19–43.

Baron, Dennis. *The English-Only Question: An Official Language for Americans?* New Haven, CT: Yale University Press, 1990.

Baud, Michael. "Imagining the Other: Michael Taussig on Mimesis, Colonialism, and Identity." *Critique of Anthropology* 17.1 (1997): 103–12.

Baugh, John. *Beyond Ebonics: Linguistic Pride and Racial Prejudice*. New York: Oxford University Press, 2000.

Baumann, Gerd. *The Multicultural Riddle: Rethinking National, Ethnic, and Religious Identities.* New York: Routledge, 1999.

Baynton, Douglas C. *Defectives in the Land: Disability and Immigration in the Age of Eugenics.* Chicago: University of Chicago Press, 2016.

Bender, Thomas. *A Nation Among Nations: America's Place in World History*. New York: Hill and Wang, 2006.

Benjamin, Jules R. *The United States and the Origins of the Cuban Revolution: An Empire of Liberty in an Age of National Liberation*. Princeton, NJ: Princeton University Press, 1990.

Berg, Jerome. *On the Shortwaves, 1923–1945: Broadcast Listening in the Pioneer Days of Radio*. Jefferson, NC: McFarland and Company, 1999.

Biesanz, John and Mavis Biesanz. *The People of Panama*. New York: Columbia University Press, 1955.

Birch, Barbara. "Ebonics: The Debate Which Never Happened." *Ethnic Studies Review* 22.1 (1999): 44–55.

Birn, Anne-Emanuelle. "Public Health or Public Menace? The Rockefeller Foundation and Public Health in Mexico, 1920–1950." *Voluntas: International Journal of Voluntary and Nonprofit Organizations* 7.1 (1996): 35–56.

Blondheim, Menahem. *News over the Wires: The Telegraph and the Flow of Public Information in America, 1844–1897*. Cambridge, MA: Harvard University Press, 1994.

Blower, Brooke L. "From Isolationism to Neutrality: A New American Framework for Understanding American Political Culture, 1919–1941." *Diplomatic History* 38.2 (2014): 345–76.

Bonfiglio, Thomas Paul. *Race and the Rise of the Standard American*. New York: De Gruyter, 2002.

Bourgois, Philippe. "One Hundred Years of United Fruit Company Letters." *Banana Wars: Power, Production, and History in the Americas*, edited by Steve Striffler and Mark Moberg, 103–44. Raleigh-Durham, NC: Duke University Press, 2003.

Brasseaux, Ryan André. *Cajun Breakdown: The Emergence of an American-Made Music.* New York: Oxford University Press, 2009.

Brison, Jeffrey D. *Rockefeller, Carnegie, and Canada American Philanthropy and the Arts and Letters in Canada*. Montréal: McGill-Queen's University Press, 2005.

Britton, John A. "'The Confusion Provoked by Instantaneous Discussion': The New International Communications Network and the Chilean Crisis of 1891–1892 in the United States." *Technology and Culture* 48.4 (2007): 729–57.

Brock, Pope. *Charlatan: America's Most Dangerous Huckster, the Man Who Pursued Him, and the Age of Flimflam.* New York: Crown Publishers, 2008.

Bronfman, Alejandra. *Isles of Noise: Sonic Media in the Caribbean*. Chapel Hill: University of North Carolina Press, 2016.

Brown, Becky. "The Development of a Louisiana French Norm." *French and Creole in Louisiana*, edited by Arthur Valdman, 215–35, New York: Plenum Press, 1997.

Brown, Stuart. "Basic English." *Encyclopedia of Rhetoric and Composition: Communication from Ancient Times to the Information Age*, edited by Theresa Enos, 68. New York: Routledge, 1996.

Brückner, Martin. "Lessons in Geography: Maps, Spellers, and Other Grammars of Nationalism in the Early Republic." *American Quarterly* 51.2 (1999): 311–43.

Brundage, W. Fitzhugh. "Le Reveil de la Louisiane: Memory and Acadian Identity." *Where These Memories Grow: History, Memory, and Southern Identity*, edited by W. Fitzhugh Brundage, 271–98. Chapel Hill: University of North Carolina Press, 2000.

———. "Memory and Acadian Identity, 1920–1960: Susan Evangeline Walker Anding, Dudley LeBlanc, and Louise Olivier, or the Pursuit of Authenticity." *Acadians and Cajuns: The Politics and Culture of French Minorities in North America*, edited by Ursula Mathis-Moser and Günter Bischof, 55–70. Innsbruck, Austria: Innsbruck University Press, 2009.

Burke, Peter J. "Relationships Among Multiple Identities." *Advances in Identity Theory and Research*, edited by Peter J. Burke, Timothy J. Owens, Richard T. Sherpe, and Peggy Thoits, 195–214. New York: Kluwer Academic/Plenum Publishers, 2003.

Buxton, William J. "John Marshall and the Humanities in Europe: Shifting Patterns of Rockefeller Foundation Support." *Minerva* 41.2 (2003): 133–53.

Canclini, Nestor García. "North Americans or Latin Americans: The Redefinition of Mexican Identity and the Free Trade Agreements." *Mass Media and Free Trade: NAFTA and the Cultural Industries*, edited by Emile G. McAnany and Kenton T. Wilkinson, 142–56. Austin: University of Texas Press, 1996.

Carson, Gerald. *The Roguish World of Doctor Brinkley*. New York: Reinhart and Company, 1960.

Casillas, Dolores Inés. *Sounds of Belonging: U.S. Spanish-Language Radio and Public Advocacy.* New York: New York University Press, 2014.

Castro, J. Justin. *Radio in Revolution: Wireless Technology and State Power in Mexico, 1897–1938*. Lincoln: University of Nebraska Press, 2016.

Cavanaugh, Jillian R. "Accent Matters: Material Consequences of Sounding Local in Northern Italy." *Language and Communications* 25 (2005): 127–48.

Chester, Edward W. *Sectionalism, Politics, and American Diplomacy*. Metuchen, NJ: Scarecrow Press, 1975.

Cinelli, Matteo, Gianmarco De Francisci Morales, Alessandro Galeazzi, Walter Quattrociocchi, and Michele Starnini. "The Echo Chamber Effect on Social Media." *Proceedings of the National Academy of Sciences* 118.9 (2021): https://doi.org/10.1073/pnas.2023301118.

Cochran, Sherman. *Big Business in China: Sino-Foreign Rivalry in the Cigarette Industry, 1890–1930*. Cambridge, MA: Harvard University Press, 1980.

Cohen, Lizabeth. *Making a New Deal: Industrial Workers in Chicago, 1919–1939*. New York: Cambridge University Press, 1990.

Cohen, Warren I. *The Asian American Century*. Cambridge, MA: Harvard University Press, 2001.

Colby, Jason M. *The Business of Empire: United Fruit, Race, and U.S. Expansion in Central America*. Ithaca, NY: Cornell University Press, 2011.

Coleman, Simon and Peter Collins. "Introduction: Ambiguous Attachments: Religion, Identity, and Nation." *Religion, Identity, and Change: Perspectives on Global Transforma-*

tions, edited by Simon Coleman and Peter Collins, 1–25. Burlington, VT: Ashgate, 2004.

Cortina, Regina. "Introduction." *The Education of Indigenous Citizens*, edited by Regina Cortina, 1–18. Tonawanda, NY: Multilingual Matters, 2014.

Craig, Douglas B. *Fireside Politics: Radio and Political Culture in the United States, 1920–1940*. Baltimore: Johns Hopkins University Press, 2000.

Cull, Nicholas. *Selling War: The British Propaganda Campaign Against American "Neutrality" in World War II*. New York: Oxford University Press, 1995.

Danbom, David B. *Born in the Country: A History of Rural America* (2nd ed.). Baltimore: Johns Hopkins University Press, 2006.

Daniel, Cletus E. *Bitter Harvest: A History of California Farmworkers, 1870–1941*. Ithaca, NY: Cornell University Press, 1981.

Daniels, Roger. *Coming to America: A History of Immigration and Ethnicity in American Life*. New York: Harper Collins, 1990.

DeConde, Alexander. *Herbert Hoover's Latin American Policy*. Palo Alto, CA: Stanford University Press, 1951.

De la Garde, Roger. "There Goes the Neighborhood: Montréal's Free Television Market and Free Trade." *Mass Media and Free Trade: NAFTA and the Cultural Industries*, edited by Emile G. McAnany and Kenton T. Wilkinson, 242–78. Austin: University of Texas Press, 1996.

Dewitt, Howard A. "Hiram Johnson and Early New Deal Diplomacy, 1933–34." *California Historical Quarterly* 53.4 (1974): 377–86.

Dewitt, Mark F. "From Chanky-Chank to Yankee Chanks: The Cajun Accordion as Identity Symbol." *The Accordion in the Americas: Klezmer, Polka, Tango, Zydeco, and More!*, edited by Helena Simonett, 44–65. Urbana: University of Illinois Press, 2012.

Doerksen, Clifford. *American Babel: Rogue Radio Broadcasters of the Jazz Age*. Philadelphia: University of Pennsylvania Press, 2005.

Doldor, Elena and Doyin Atewologun. "Why Work It When You Can Dodge It? Identity Responses to Ethnic Stigma Among Professionals." *Human Relations* 74.6 (2021): 892–921.

Dominguez, Virginia R. *White by Definition: Social Classification in Creole Louisiana*. New Brunswick, NJ: Rutgers University Press, 1986.

Douglas, Susan J. *Inventing American Broadcasting, 1899–1922*. Baltimore: Johns Hopkins University Press, 1989.

———. *Listening In: Radio and the American Imagination, from Amos 'n' Andy and Edward R. Murrow to Wolfman Jack and Howard Stern*. New York: Times Books, 1999.

———. "The Turn Within: The Irony of Technology in a Globalized World." *American Quarterly* 58.3 (2006): 619–38.

Dubois, Sylvie, Emilie Gagnet Leumas, and Malcolm Richardson. *Speaking French in Louisiana, 1720–1955: Linguistic Practices of the Catholic Church*. Baton Rouge: Louisiana State University Press, 2018.

Dubois, Sylvie and Megan Melançon. "Cajun Is Dead—Long Live Cajun: Shifting from a Linguistic to a Cultural Community." *Journal of Sociolinguistics* 1.1 (1997): 63–93.

Dwyer, John F. "Diplomatic Weapons of the Weak: Mexican Policymaking during the U.S.-Mexican Agrarian Dispute, 1934–1941." *Diplomatic History* 26.3 (2002): 375–95.

Ealy, Lawrence O. *The Republic of Panama in World Affairs, 1903–1950*. Philadelphia: University of Pennsylvania Press, 1951.

Eckes, Jr., Alfred E. *Opening America's Market: U.S. Foreign Trade Policy Since 1776*. Chapel Hill: University of North Carolina Press, 1995.

Edwards, John. *Language and Identity: An Introduction*. New York: Cambridge University Press, 2009.

Egri, Carolyn P., David A. Ralston, Cheryl S. Murray, and Joel D. Nicholson. "Managers in the NAFTA Countries: A Cross-Cultural Comparison of Attitudes Toward Upward Influence Strategies." *Journal of International Management* 6.2 (2000): 149–71.

Eidsheim, Nina Sun. "Marian Anderson and 'Sonic Blackness' in American Opera." *American Quarterly* 63.3 (2011): 641–71.

Etulain, Richard W. and Michael P. Malone. *The American West: A Modern History, 1900 to the Present*, 2nd ed. Lincoln: University of Nebraska Press, 2007.

Finn, Seth. "Office of Radio Research." *The Concise Encyclopedia of American Radio*, edited by Christopher Sterling and Cary O'Dell, 534–36. New York: Routledge, 2010.

Fleuriet, K. Jill. *Rhetoric and Reality on the U.S.-Mexican Border: Place, Politics, and Home*. New York: Palgrave Macmillan, 2021.

Foust, James C. *Big Voices of the Air: The Battle over Clear Channel Radio*. Ames: Iowa State University Press, 2000.

Fowler, Gene and Bill Crawford. *Border Radio: Quacks, Yodelers, Pitchmen, Psychics, and Other Amazing Broadcasters of the American Airwaves*, rev. ed. Austin: University of Texas Press, 2002.

Friedman, Max Paul. "The Good Neighbor Policy." *Oxford Research Encyclopedia of Latin American History* (2018): https://doi.org/10.1093/acrefore/9780199366439.013.222.

Frost, Gary. *Early FM Radio: Incremental Technology in Twentieth-Century America*. Baltimore: Johns Hopkins University Press, 2010.

Fry, Joseph A. "Place Matters: Domestic Regionalism and the Formation of American Foreign Policy." *Diplomatic History* 36.3 (2012): 451–82.

Gagné, Gilbert. "Cultural Sovereignty, Identity, and North American Integration: On the Relevance of the U.S.-Canada-Quebec Border." *Quebec Studies* 36.1 (2003): 29–49.

———. "The Treatment of Cultural Products and the Cultural Exemption Clause." *NAFTA 2.0: From the First NAFTA to the United-States-Mexico-Canada Agreement*, edited by Gilbert Gagné and Michèle Rioux, 243–60. New York: Palgrave MacMillan, 2022.

Galperin, Hernan. "Cultural Industries Policy in Regional Trade Agreements: The Cases of NAFTA, the European Union and MERCOSUR." *Media, Culture, and Society* 21.5 (1999): 627–48.

Gellman, Irwin F. *Good Neighbor Diplomacy: United States Policies in Latin America, 1933–1945*. Baltimore: Johns Hopkins University Press, 1979.

Gerstle, Gary. *American Crucible: Race and Nation in the Twentieth Century*. Princeton, NJ: Princeton University Press, 2001.

Gienow-Hecht, Jessica. *Sound Diplomacy: Music and Emotions in Transatlantic Relations, 1850–1920*. Chicago: University of Chicago Press, 2009.

Gilderhus, Mark T. *The Second Century: U.S.-Latin American Relations since 1889*. Wilmington, DE: Scholarly Resources, 2000.

Godfrey, Dan and Alton Franklin. "Extension Programs at the 1890 Land-Grant Institutions." *A Century of Service: Land-Grant Colleges and Universities, 1890–1990*, edited by Ralph D. Christy and Lionel Williamson, 59–68. New Brunswick, NJ: Transaction Publishers, 2012.

Godfrey, Donald G. and Louise M. Benjamin. "Radio Legislation's Quiet Backstage Negotiator: Wallace H. White, Jr." *Journal of Radio Studies* 10.1 (2003): 93–103.

Godfried, Nathan. "Struggling over Politics and Culture: Organized Labor and Radio Station WEVD during the 1930s." *Labor History* 42.4 (2001): 347–69.

———. *WCFL: Chicago's Voice of Labor, 1926–1978*. Urbana: University of Illinois Press, 1997.

Greene, Julia. *The Canal Builders: The Making of America's Empire at the Panama Canal*. New York: Penguin Press, 2009.

Greenfeld, Liah. "The Origins and Nature of American Nationalism in Comparative Perspective." *The American Nation—National Identity—Nationalism*, edited by Knud Krakau, 19–52. New Brunswick, NJ: LIT Vertag, 1997.

Gvirtz, Silvina, Jason Beech, and Angela Oria. "Schooling in Argentina." *Going to School in Latin America*, edited by Silvina Gvirtz and Jason Beech, 5–33. Westport, CT: Greenwood Press, 2008.

Hall, Peter Dobkin. "Business, Philanthropy, and Education in the United States." *Theory into Practice* 33.4 (1994): 211–17.

Hamilton, Sheila. "Panamanian Politics and Panama's Relationship with the United States Leading up to the Hull Alfaro Treaty." MA thesis, University of Victoria, 2009.

Harding, Robert. "The Military Foundations of Panamanian Politics: From the National Police to the PRD and Beyond." PhD dissertation: University of Miami, 1998.

Hayes, Joy Elizabeth. *Radio Nation: Communication, Popular Culture and Nationalism in Mexico, 1920–1950*. Tucson: University of Arizona Press, 2000.

Headrick, Daniel R. *The Invisible Weapon, Telecommunications and International Politics, 1851–1945*. New York: Oxford University Press, 1991.

———. "Shortwave Radio and Its Impact on International Telecommunications Between the Wars." *History and Technology* 11.1 (1994): 21–32.

Heffernan, Michael. "The Cartography of the Fourth Estate: Mapping the New Imperialism in British and French Newspapers, 1875–1925." *The Imperial Map: Cartography and the Mastery of Empire*, edited by James R. Ackerman, 261–99. Chicago: University of Chicago Press, 2009.

Heil, Jr., Alan L. *Voice of America: A History*. New York: Columbia University Press, 2003.

Henry, Jacques. "The Louisiana French Movement: Actors and Actions in Social Change." *French and Creole in Louisiana*, edited by Arthur Valdman, 183–213. New York: Plenum Press, 1997.

Herb, Guntram H. "National Identity and Territory." *Nested Identities: Nationalism, Territory and Scale*, edited by Guntram H. Herb and David H. Kaplan, 9–30. Lanham, MD: Rowman and Littlefield, 1999.

——. *Under the Map of Germany: Nationalism and Propaganda, 1918–1945*. New York: Routledge, 1997.

Hernandez-Galano, Yamilet and Lyding R. Rodriquez Fuentes. "Schooling in Cuba." *Going to School in Latin America*, edited by Silvina Gvirtz and Jason Beech, 143–77. Westport, CT: Greenwood Press, 2008.

Hilliard, Robert and Michael Keith. *The Quieted Voice: The Rise and Demise of Localism in American Radio*. Carbondale: University of Southern Illinois Press, 2005.

Hilliker, Jim. "KFI." *The Encyclopedia of Radio*, edited by Christopher Sterling, 815–17. New York: Taylor and Francis, 2004.

Hills, Jill. *The Struggle for Control of Global Communication: The Formative Century*. Urbana: University of Illinois Press, 2002.

Hilmes, Michele. *Hollywood and Broadcasting: From Radio to Cable.* Urbana: University of Illinois Press, 1990.

——. *Network Nations: A Transnational History of British and American Broadcasting*. New York: Routledge, 2012.

——. "The New Vehicle of Nationalism: Radio Goes to War." *The Oxford Handbook of Propaganda Studies*, edited by Jonathan Auerbach and Russ Castronovo, 201–18. New York, Oxford University Press, 2014.

——. *Radio Voices: American Broadcasting, 1922–1952*. Minneapolis: University of Minnesota Press, 1997.

Hogan, J. Michael. *The Panama Canal in American Politics: Domestic Advocacy and the Evolution of Policy*. Carbondale: Southern Illinois University Press, 1986.

Hoganson, Kristin. "Stuff It: Domestic Consumption and the Americanization of the World Paradigm." *Diplomatic History* 30.4 (2006): 571–94.

Horsman, Reginald. *Race and Manifest Destiny: The Origins of American Racial Anglo-Saxonism.* Cambridge, MA: Harvard University Press, 1981.

Horten, Gerd. *Radio Goes to War: The Cultural Politics of Propaganda during World War II.* Berkeley: University of California Press, 2002.

Hudley, Anne H. Charity. "Language and Racialization." *The Oxford Handbook of Language and Society*, edited by Ofelia García, Nelson Flores, and Massimiliano Spotti, 381–402. New York: Oxford University Press, 2017.

Ide, Yoshimitsu. "The Significance of Richard Hathaway Edmonds and His *Manufacturers' Record* in the New South." PhD dissertation: University of Florida, 1959.

Jacobson, Matthew Frye. *Barbarian Virtues: The United States Encounters Foreign Peoples and Home and Abroad, 1876–1917*. New York: Hill and Wang, 2000.

——. *Whiteness of a Different Color: European Immigrants and the Alchemy of Race*. Cambridge, MA: Harvard University Press, 1998.

Joseph, John E. *Language and Identity: National, Ethnic, Religious*. New York: Palgrave Macmillan, 2004.

Judt, Tony and Denis Lacorne. "The Politics of Language." *Language, Nation, and State: Identity Politics in a Multilingual Age*, edited by Tony Judt and Denis Lacorne, 1–16. New York: Palgrave MacMillan, 2004.

Juhnke, Eric S. *Quacks and Crusaders: The Fabulous Careers of John Brinkley, Norman Baker, and Harry Hoxsey*. Lawrence: University Press of Kansas, 2002.

Kaestle, Carl F. *Pillars of the Republic: Common Schools and American Society, 1780–1860*. New York: Harper Collins, 1983.

Karush, Matthew B. *Culture of Class: Radio and Cinema in the Making of a Divided Argentina, 1920–1946*. Durham, NC: Duke University Press, 2012.

Kelman, Ari. *Station Identification: A Cultural History of Yiddish Radio in the United States*. Berkeley: University of California Press, 2009.

Kermes, Stephanie. *Creating an American Identity: New England, 1789–1825*. New York: Palgrave Macmillan, 2008.

Kern, Stephen. *The Culture of Time and Space, 1880–1918*. Cambridge, MA: Harvard University Press, 1983.

Kirby, Dianne. "Divinely Sanctioned: The Anglo-American Cold War Alliance and the Defence of Western Civilization and Christianity, 1945–48." *Journal of Contemporary History* 35.3 (2000): 385–412.

Kirschner, Don S. *City and Country: Rural Responses to Urbanization in the 1920s*. Westport, CT: Greenwood Press, 1970.

Kline, Ronald R. *Consumers in the Country: Technology and Social Change in Rural America*. Baltimore: Johns Hopkins University Press, 2000.

Klingler, Thomas A. "How Much Acadian Is There in Cajun?" *Acadians and Cajuns: The Politics and Culture of French Minorities in North America*, edited by Ursula Mathis-Moser and Günther Bischof, 91–106. Innsbruck, Austria: University of Innsbruck Press, 2009.

Koeneke, Rodney. *Empires of the Mind: I.A. Richards and Basic English in China*. Stanford, CA: Stanford University Press, 2004.

Krysko, Michael A. *American Radio in China: International Encounters with Technology and Communications, 1919–41*. New York: Palgrave MacMillan, 2011.

Kuisel, Richard. *Seducing the French: The Dilemma of Americanization*. Berkeley: University of California Press, 1993.

——. *The French Way: How France Embraced and Rejected American Values and Power*. Princeton, NJ: Princeton University Press, 2012.

Lacey, Kate. *Listening Publics: The Politics and Experience of Listening in the Media Age*. Malden, MA: Polity Press, 2013.

LaFeber, Walter. *The American Age: U.S. Foreign Policy at Home and Abroad, 1750 to the Present* (2nd ed.). New York: W.W. Norton, 1996.

——. *The Panama Canal: The Crisis in Historical Perspective*, updated edition. New York: Oxford University Press, 1989.

Langer, Nils. *Linguistic Purism in Action: How Auxiliary Tun Was Stigmatized in Early New High German*. New York: Walter de Gruyter, 2001.

Langley, Lester. "Negotiating New Treaties with Panama: 1936." *Hispanic American Historical Review* 48.2 (1968): 220–33.

———. "The World Crisis and the Good Neighbor Policy in Panama, 1936–41." *The Americas* 24.2 (1967): 137–52.

Lebovics, Herman. *True France: The Wars over Cultural Identity, 1900–1945*. Ithaca, NY: Cornell University Press, 1992.

Lee, R. Alton. *The Bizarre Careers of John R. Brinkley*. Lexington: The University Press of Kentucky, 2002.

Leeman, Jennifer. "Categorizing Latinos in the History of the US Census: The Official Racialization of Spanish." *A Political History of Spanish: The Making of a Language*, edited by José Del Valle, 305–23. New York: Cambridge University Press, 2013.

Lenthall, Bruce. *Radio's America: The Great Depression and the Rise of Modern Mass Culture.* Chicago: University of Chicago Press, 2007.

Leonard, Thomas. *Historical Dictionary of Panama.* Lanham, MD: Rowman and Littlefield, 2015.

———. "The United States and Panama: Negotiating the Aborted 1926 Treaty." *Mid America: An Historical Review* 61.3 (1979): 189–203.

Lindemann, Stephanie. "Who Speaks 'Broken English'? US Undergraduate Perceptions of Non-Native English." *International Journal of Applied Linguistics* 15.2 (2005): 187–212.

Lipsitz, George. "Land of a Thousand Dances: Youth, Minorities, and the Rise of Rock and Roll." *Recasting America: Culture and Politics in the Age of Cold War*, edited by Lary May, 267–84. Chicago: University of Chicago Press, 1989.

Lommers, Suzanne. *Europe—On Air: Interwar Projects for Radio Broadcasting.* Amsterdam: Amsterdam University Press, 2012.

Lopez, Luis Enrique. "Indigenous Intercultural Bilingual Education in Latin America: Widening Gaps between Policy and Practice." *The Education of Indigenous Citizens*, edited by Regina Cortina, 19–49. Tonawanda, NY: Multilingual Matters, 2014.

Loviglio, Jason. "Moments of Danger: The Struggle for Community-based Public Radio in Baltimore." *Journal of Alternative and Community Media* 4.4 (2019): 37–50.

———. *Radio's Intimate Voice: Network Broadcasting and Mass Mediated Democracy.* Baltimore: Johns Hopkins University Press, 2005.

Lowenthal, Abraham F. *Global California: Rising to the Cosmopolitan Challenge*. Stanford, CA: Stanford University Press, 2009.

Major, John. *Prize Possession: The United States and the Panama Canal, 1903–1979.* New York: Cambridge University Press, 1993.

Mamakos, Michalis and Eli J. Finkel, "The Social Media Discourse of Engaged Partisans Is Toxic Even When Politics Are Irrelevant." *PNAS Nexus* 2.10 (2023): https://academic.oup.com/pnasnexus/article/2/10/pgad325/7293179.

Marsh, Steve. "The Special Relationship and the Anglo-Iranian Oil Crisis, 1950–4." *Review of International Studies* 24.4 (1998): 529–44.

Martin, James W. *Banana Cowboys: The United Fruit Company and the Culture of Corporate Colonialism.* Albuquerque: University of New Mexico Press, 2018.

May, Lary. *Screening Out the Past: The Birth of Mass Culture and the Motion Picture Industry*, rev. ed. Chicago: University of Chicago Press, 1983.

Mayes, McKinley. "Status of Agricultural Research Programs at 1890 Land-Grant Institutions and Tuskegee University." *A Century of Service: Land-Grant Colleges and Universities, 1890–1990*, edited by Ralph D. Christy and Lionel Williamson, 53–58. New Brunswick, NJ: Transaction Publishers, 2012.

McAnany, Emile G. and Kenton T. Wilkinson. "Introduction." *Mass Media and Free Trade: NAFTA and the Cultural Industries*, edited by Emile G. McAnany and Kenton T. Wilkinson, 3–29. Austin: University of Texas Press, 1996.

McChesney, Robert W. *Telecommunications, Mass Media, and Democracy: The Battle for Control of U.S. Broadcasting, 1928–1935*. New York: Oxford University Press, 1993.

McCrea, Heather L. "Treating Delinquent and Feebleminded Juveniles at the Beloit Industrial School for Girls in Early Twentieth-Century Kansas." *Journal of the History of Medicine and Allied Sciences*, 79.3 (2024): 193–211.

McCreery, David. "Wireless Empire: The United States and Radio Communications in Central America and the Caribbean, 1904–1926." *South Eastern Latin Americanist* 37.1 (1993): 23–41.

McPherson, Alan. "*Antiyanquismo*: Nascent Scholarship, Ancient Sentiments." *Anti-Americanism in Latin America and the Caribbean*, edited by Alan McPherson, 1–34. New York: Berghahn Books, 2006.

——. "Anti-Americanism in Latin America." *Anti-Americanism: History, Causes, and Themes* (vol. 3: Comparative Perspectives), edited by Brendon O'Connor, 77–102. Westport, CT: Greenwood World Publishing, 2007.

——. "Herbert Hoover, Occupation Withdrawal, and the Good Neighbor Policy." *Presidential Studies Quarterly* 44.4 (2014): 623–39.

Melis, Rolando Poblete. "Schooling in Chile." *Going to School in Latin America*, edited by Silvina Gvirtz and Jason Beech, 77–95. Westport, CT: Greenwood Press, 2008.

Messere, Fritz. "The Davis Amendments and the Federal Radio Act of 1927: External Pressures in Policy Making." *Transmitting the Past: Historical and Cultural Perspectives on Broadcasting*, edited J. Emmett Winn and Susan L. Brinson, 34–68. Tuscaloosa: University of Alabama Press, 2005.

Millar, Robert McColl. *Language, Nation and Power*. New York: Palgrave MacMillan, 2005.

Miller, Edward D. *Emergency Broadcasting and 1930s American Radio*. Philadelphia: Temple University Press, 2003.

Moreno, Julio. *Yankee Don't Go Home! Mexican Nationalism, American Business Culture, and the Shaping of Modern Mexico, 1920–1950*. Chapel Hill: University of North Carolina Press, 2003.

Morrice, Linda. "Refugees in Higher Education: Boundaries of Belonging and Recognition, Stigma and Exclusion." *International Journal of Lifelong Education* 32.5 (2013): 652–68.

Munro, Dana Gardner. *Intervention and Dollar Diplomacy in the Caribbean, 1900–1921*. Princeton, NJ: Princeton University Press, 1964.

———. *The United States and the Caribbean Republics, 1921–1933*. Princeton, NJ: Princeton University Press, 1974.

Nash, Gary. *The American West Transformed: The Impact of the Second World War.* Bloomington: Indiana University Press, 1985.

Nathan, Andrew J. *Chinese Democracy*. Berkeley: University of California Press, 1986.

Neilson, Keith. "Perception and Posture in Anglo-American Relations: The Legacy of the Simon-Stimson Affair, 1932–1941." *International History Review* 29.2 (2007): 313–37.

Neulander, Joelle. *Programming National Identity: The Culture of Radio in 1930s France.* Baton Rouge: Louisiana State University Press, 2009.

Ngai, Mae. *Impossible Subjects: Illegal Aliens and the Making of Modern America*, rev. ed. Berkeley: University of California Press, 2014.

Nichols, Christopher McKnight. *Promise and Peril: America at the Dawn of a Global Age*. Cambridge, MA: Harvard University Press, 2011.

Nickles, David Paull. *Under the Wire: How the Telegraph Changed Diplomacy*. Cambridge, MA: Harvard University Press, 2009.

Noel, Linda. *Debating American Identity: Southwest Statehood and Mexican Immigration*. Tucson: University of Arizona Press, 2014.

Nye, David E. *Technology Matters: Questions to Live With.* Cambridge, MA: MIT Press, 2006.

Obregon, Javier Sáenz and Oscar Saldarriaga Vélez. "Schooling in Colombia." *Going to School in Latin America*, edited by Silvina Gvirtz and Jason Beech, 97–118. Westport, CT: Greenwood Press, 2008.

Offner, Arnold. "'Another Such Victory': President Truman, American Foreign Policy, and the Cold War." *Diplomatic History* 23.2 (1999): 127–55.

Parsons, Patrick R. "The Lost Doctrine: *Suggestion Theory* in Early Media Effects Research." *Journalism and Communications Monographs* 23.2 (2021): 80–138.

Paterson, Thomas G. *Contesting Castro: The United States and the Triumph of the Cuban Revolution*. New York: Oxford University Press, 1994.

Perez, Jr., Louis A. *Cuba Under the Platt Amendment, 1902–1934*. Pittsburgh: University of Pittsburgh Press, 1986.

———. *On Becoming Cuban: Identity, Nationality, and Culture*. Chapel Hill: University of North Carolina Press, 2001.

Perry, Theresa and Lisa Delpit, eds. *The Real Ebonics Debate: Power, Language, and the Education of African-American Children*. Boston: Beacon Press, 1998.

Peters, John Durham. *Speaking Into the Air: A History of the Idea of Communication*. Chicago: University of Chicago Press, 1999.

Pomeroy, Earl. *The Pacific Slope: A History of California, Oregon, Washington, Idaho, Utah, and Nevada*. Lincoln: University of Nebraska Press, 1991.

Pooley, Jefferson. "The New History of Mass Communication Research." *The History of Media and Communication Research: Contested Memories*, edited by David W. Park and Jefferson Pooley, 43–70. New York: Peter Lang, 2008.

Potter, Simon J. *Wireless Internationalism and Distant Listening: Britain, Propaganda, and the Invention of Global Radio, 1920–1939*. New York: Oxford University Press, 2020.

Raat, W. Dirk and Michael M. Brescia, *Mexico and the United States: Ambivalent Vistas*, 4th ed. Athens: University of Georgia Press, 2010.

Rantanen, Tehri. *The Media and Globalization*. Thousand Oaks, CA: SAGE, 2005.

Razlogova, Elena. *The Listener's Voice: Early Radio and the American Public*. Philadelphia: University of Pennsylvania Press, 2011.

Reese, William J. *History, Education, and the Schools*. New York: Palgrave MacMillan, 2007.

Resnick, Philip. *The Labyrinth of North American Identities*. Toronto: University of Toronto Press, 2012.

Reynolds, David. "Rethinking Anglo-American Relations." *International Affairs* 65.1 (1988): 89–111.

Richeri, Giuseppe. "Global Film Market, Regional Problems." *Global Media and China* 1.4 (2016): 312–30.

Rivero, Yeidy M. *Broadcasting Modernity: Cuban Commercial Television, 1950–1960*. Durham, NC: Duke University Press, 2015.

Robles, Sonia. *Mexican Waves: Radio Broadcasting Along Mexico's Northern Border, 1930–1950*. Tucson: University of Arizona Press, 2019.

Rosa, Jonathan and Nelson Flores, "Unsettling Race and Language: Toward a Raciolinguistic Perspective." *Language in Society* 46.5 (2017): 621–47.

Rosenberg, Emily. "Transnational Currents in a Shrinking World." *A World Connecting, 1870–1945*, edited by Emily Rosenberg, 815–998. Cambridge, MA: Belknap Press of Harvard University Press, 2012.

———. *Spreading the American Dream: American Economic and Cultural Expansion, 1890–1945*. New York: Hill and Wang, 1983.

Rumer, Thomas A. *The American Legion: An Official History*. New York: M. Evans, 1990.

Russo, Alexander. *Points on the Dial: Golden Age Radio Beyond the Networks*. Durham, NC: Duke University Press, 2010.

Saettler, Paul. *The Evolution of American Education Technology*. Greenwich, CT: Information Age Publishing, 2004.

Salwen, Michael B. "Lemmon, Walter S." *Historical Dictionary of Political Communication*, edited by Guido Hermann Stempel, 79. Westport, CT: Greenwood Press, 1999.

———. *Radio and Television in Cuba: The Pre-Castro Era*. Ames: Iowa State University Press, 1994.

Sammond, Nicholas. *Babes in Tomorrowland: Walt Disney and the Making of the American Child, 1930–1960*. Durham, NC: Duke University Press, 2005.

Savage, Barbara Dianne. *Radio, War, and the Politics of Race, 1938–1948*. Chapel Hill: University of North Carolina Press, 1999.

Scales, Rebecca. *Radio and the Politics of Sound in Interwar France, 1921–1939*. New York: Cambridge University Press, 2016.

Schumacher, Frank. "Embedded Empire: The United States and Colonialism." *Journal of Modern European History* 14.2 (2016): 202–24.

Schwoch, James. *The American Radio Industry and Its Latin American Activities, 1900–1939*. Urbana: University of Illinois Press, 1989.

———. *Global TV: New Media and the Cold War, 1946–1969*. Urbana: University of Illinois Press, 2009.

Seidelhofer, Barbara. *Understanding English as a Lingua Franca*. New York: Oxford University Press, 2011.

Self, Robert. "Perception and Posture in Anglo-American Relations: The War Debt Controversy in the 'Official Mind,' 1919–1940." *International History Review* 29.2 (2007): 282–312.

Shepperd, Josh. "Infrastructure in the Air: The Office of Education and the Development of Public Broadcasting in the United States." *Critical Studies in Media Communications* 31.3 (August 2014): 230–43.

———. *Shadow of the New Deal: The Victory of Public Broadcasting*. Urbana: University of Illinois Press, 2023.

Shideler, James H. "'Flappers and Philosophers,' and Farmers: Rural-Urban Tensions of the Twenties." *Agricultural History* 47.4 (1973): 283–99.

Sinclair, John. "Culture and Trade: Some Theoretical and Practical Considerations." *Mass Media and Free Trade: NAFTA and the Cultural Industries*, edited by Emile G. McAnany and Kenton T. Wilkinson, 30–62. Austin: University of Texas Press, 1996.

Slay, Holly S. and Delmonize A. Smith, "Professional Identity Construction: Using Narrative to Understand the Negotiation of Professional and Stigmatized Cultural Identities." *Human Relations* 64.1 (2011): 85–107.

Slotten, Hugh R. *Beyond Sputnik and the Space Race: The Origins of Global Satellite Communications*. Baltimore: Johns Hopkins University Press, 2022.

———. *Radio and Television Regulation: Broadcasting Technology in the United States, 1920–1960*. Baltimore: Johns Hopkins University Press, 2000.

———. *Radio's Hidden Voice: The Origins of Public Broadcasting in the United States*. Urbana: University of Illinois Press, 2009.

Smith, Merritt Roe and Leo Marx, eds. *Does Technology Drive History? The Dilemma of Technological Determinism*. Cambridge, MA: MIT Press, 1994.

Smith, Peter. *Talons of the Eagle: Dynamics of U.S.-Latin American Relations* (2nd ed.). New York: Oxford University Press, 1999.

Smulyan, Jeffrey H. "Power to Some People: The FCC's Clear Channel Allocation Policy." *Southern California Law Review* 44.3 (1971): 811–47.

Smulyan, Susan. *Selling Radio: The Commercialization of American Broadcasting, 1920–1934*. Washington, DC: Smithsonian Institution Press, 1994.

Staff, Hector H. *Historia y testimonios de la radiodifusion en Panamá*. Universidad de Panamá, Facultad de Comunicación Social, Escuela de Radiodifusión, 1985.

———. *La voz del barú y otros aspectos de radiodifusión y soberanía*. Panama: Mes de la Patria, 1988.

Stamm, Michael. "Paul Lazarsfeld's *Radio and the Printed Page*: A Critical Reappraisal," *American Journalism* 27.4 (2010): 37–58.

Stanton, Andrea. *"This Is Jerusalem Calling": State Radio in Mandate Palestine.* Austin: University of Texas Press, 2013.

Sterling, Christopher H. "Clear Channel Stations: Powerful Major Market Radio Stations." *The Encyclopedia of Radio*, edited by Christopher Sterling, 342–44. New York: Fitzroy Dearborn, 2004.

———. "North American Regional Broadcasting Agreement: Sharing Frequencies Among the United States, Canada, Mexico, and the Caribbean." *The Concise Encyclopedia of American Radio*, edited by Christopher Sterling and Cary O'Dell, 528–29. New York: Routledge, 2010.

Stoner, K. Lynn. "Militant Heroines and Consecration of the Patriarchal State." *Cuban Studies* 34 (2003): 71–96.

Stryker, Sheldon and Peter J. Burke. "The Past Present and Future of an Identity Theory." *Social Psychology Quarterly* 63.4 (2000): 284–97.

Thoits, Peggy A. "Personal Agency in Multiple Role Identities." *Advances in Identity Theory and Research*, edited by Peter J. Burke, Timothy J. Owens, Richard T. Sherpe, and Peggy Thoits, 179–94. New York: Kluwer Academic/Plenum Publishers, 2003.

Thomas, George. *Linguistic Purism.* London: Longman, 1991.

Thrower, Norman J.W. *Maps and Civilization: Cartography in Culture and Society* (2nd ed.). Chicago: University of Chicago Press, 2008.

Toyoki, Sammy and Andrew Brown. "Stigma, Identity, and Power: Managing Stigmatized Identities through Discourse." *Human Relations* 67.6 (2014): 715–37.

Tyson, Timothy B. *Radio Free Dixie: Robert F. Williams and the Roots of Black Power.* Chapel Hill: University of North Carolina Press, 1999.

Urban, Wayne J. and Jennings L. Wagoner Jr. *American Education: A History* (4th ed.). New York: Routledge, 2009.

Vaillant, Derek W. *Across the Waves: How the United States and France Shaped the International Age of Radio.* Chicago: University of Illinois Press, 2017.

———. *Sounds of Reform: Progressivism and Music in Chicago, 1873–1935.* Chapel Hill: University of North Carolina Press, 2003.

Van Rossem, Ronan. "The World System Paradigm as General Theory of Development: A Cross-National Test." *American Sociological Review* 61.3 (1996): 508–27.

Vargas, Zaragosa. *Proletarians of the North: A History of Mexican Industrial Workers in Detroit and the Midwest, 1917–1933.* Berkeley: University of California Press, 1993.

Vaughan, Mary Kay. *Cultural Politics in Revolution: Teachers, Peasants, and Schools in Mexico, 1930–1940.* Tucson: University of Arizona Press, 1997.

Vogel, Ann. "Who's Making Global Civil Society: Philanthropy and US Empire in World Society." *British Journal of Sociology* 57.4 (2006): 635–55.

Wade, Michael G. "Villainy, Virtue, and Louisiana Political Culture: Paul Herbert and the Augean Stables at LSU, 1939–1941." *Louisiana History: The Journal of the Louisiana Historical Association* 49.1 (2008): 655–82.

Walker, Jesse. *Rebels on the Air: An Alternative History of Radio in America.* New York: NYU Press, 2001.

Walker, Samuel. *Presidents and Civil Liberties from Wilson to Obama: A Story of Poor Custodians*. New York: Cambridge University Press, 2014.

Walsh, Daniel. *An Air War with Cuba: The United States Radio Campaign Against Castro.* Jefferson, NC: McFarland and Company, 2012.

Walsh, Olivia. *Linguistic Purism: Language Attitudes in France and Quebec*. Philadelphia: John Benjamins Publishing Company, 2016.

Watson, George. "The Amiable Heretic: I.A. Richards, 1893–1979." *Sewanee Review* 104.2 (1996): 248–62.

Weinstein, Brian. "Francophonie: Purism at the International Level." *The Politics of Language Purism*, edited by Björn H. Jernudd and Michael J. Shapiro, 53–80. New York: Mouton de Gruyter, 1989.

Wheatley, Steven C. "Introduction." *The Story of the Rockefeller Foundation*, by Raymond Fosdick, vii–xix. New York: Routledge, 2017.

White, Lillian C. *Pioneer and Patriot: George Cook Sweet, Commander, U.S.N., 1877–1953: A Biography.* Delray Beach, FL: The Southern Publishing Company, 1963.

Wiley, Terrence G. "Continuity and Change in the Function of Language Ideologies in the United States." *Ideology, Politics, and Language Policies: Focus on English*, edited by Thomas Ricento, 67–86. Philadelphia: John Benjamins Publishing Company, 2000.

Winford, Donald. "Ideologies of Language and Socially Realistic Linguistics." *Black Linguistics: Language, Society, and Politics in Africa and the Americas*, edited by Sinfree Makoni, Geneva Smitherman, Arnetha F. Ball, and Arthur K. Spears, 21–39. New York: Routledge Press, 2003.

Winkler, Jonathan Reed. *NEXUS: Strategic Communications and American Security in World War I.* Cambridge, MA: Harvard University Press, 2008.

Wong, Andrew D., Hsi-Yao Su, and Mie Hiramoto, "Complicating Raciolinguistics: Language, Chineseness, and the Sinophone." *Language and Communication* 76 (2021): 131–35.

Wood, Bryce. *The Making of the Good Neighbor Policy*. New York: Columbia University Press, 1961.

Wood, James. *History of International Broadcasting* (vol. 2.). London: The Institution of Electrical Engineers, 2000.

Woodward, Colin. *American Nations: A History of the Eleven Rival Regional Cultures of North America*. New York: Penguin Books, 2011.

Wright, Micah. "Unilateral Pan-Americanism: Wilsonianism and the American Occupation of Chiriquí." *Diplomacy and State Craft* 26.1 (2015): 46–64.

Yang, Daqing. *Technology of Empire: Telecommunications and Japanese Expansion in Asia, 1883–1945*. Cambridge, MA: Harvard University Press, 2010.

Zajácz, Rita. "Fragmented Imperialism: US Control over Radio in Panama, 1914–36." *International Communications Gazette* 74.1 (2012): 78–94.

Zulwaski, Ann. *Unequal Cures: Public Health and Political Change in Bolivia, 1900–1950*. Durham, NC: Duke University Press, 2007.

INDEX

Abbott and Costello, 22
Acadian: culture, 67; people, 71–72. *See also* Cajun
Acadian Handicrafts Project, 86
accents, 3, 6, 26, 109, 175
Adams, Charles Francis, 140–41
Adams, Charles M., 20, 55
Africa, 96
Alabama, 176
Alaska, 27–28
Alerta (Cuba), 160
Alfaro Jované, Ricardo, 130–31
Alfaro-Kellogg Treaty (1926), 135, 139–40
All America Cables, 140–41
Altaffer, Maurice, 44
amateur radio, 34–35, 130–31, 149, 186n44
American Civil Liberties Union, 35
American exceptionalism, 9, 91, 96. *See also* AngloSaxon exceptionalism
Americanization (of immigrants). *See* assimilation
American Legion, 35
American Medical Association, 43
American system of radio broadcasting, 7–8; development of, 15, 21–23, 26–28, 37; listeners and, 41, 50–51, 56, 60–61; public service broadcasting impacted by, 69. *See also* clear channel stations and frequencies
Amos 'n' Andy, 6, 22
Anderson, Benedict, 27
Andrews, Earl, 15–16, 28–29, 34, 37
AngloSaxon exceptionalism, 9, 91, 101–2. *See also* American exceptionalism
anti-Americanism: in Cuba, 10, 153, 156–57, 160, 171; in Latin America and the Caribbean, 111, 114, 134; in Mexico, 48–49, 111, 171–72; in Panama, 10, 118, 127–28, 132, 138–40, 148–51, 171–72, 178
anti-Semitism, 58
Arcadia High School (Louisiana), 59
Argentina, 108
Arias Madrid, Harmodio, 142
Arizona, 11, 152, 161–62, 169–70
Arlington, New Jersey, 30
Army, United States. *See* United States Army
Arosemena Barreati, Juan Demóstenes, 139, 141–42
Arosemena Guillén, Florencio Harmodio, 139
assimilation, 15, 19–20, 29–31, 35–37, 54–55, 177; Groucho Marx's views of, 34

Associated Farmers of California, 162. *See also* Farm Bureau
Associated Press, 168, 181n11
Atlanta, Georgia, 50–52, 166
Aurora, Nebraska, 41
Australia, 110
Austria, 120
Averill, Harvey, 39, 41–42

Badger, R. E., 162
Bahamas, 158, 161
Baker, Norman, 26, 46
Basic English, 9, 91, 98–105, 110–14, 172, 178
Batista, Fulgencio, 157
Bauer, William, 32
Behn, Sosthenes, 141
Belgium, 120
Bell, Walter, 32
Bender, Thomas, 4
Biden, Joseph, 177
Bielecki, Louis, 24–25
Birkhead, Leon Milton, 58
Black English. *See* Ebonics
Blake, Katherine Deveraux, 2–3
Blower, Brook, 218n24
Bocas del Toro, Panama, 121
Bolivia, 109
"border blaster" stations, 26, 57–61, 153, 178. *See also* XER
Boston, Massachusetts: United Fruit in, 121–22; World Wide Broadcasting Foundation in, 9, 90, 95, 97–98, 104
Boston Globe, 24
Boston-Panama Company, 130
Boston University, 90
boucheries, 75. *See also* Cajun: culture
Boyle, Walter, 47
Braille, 100
Brazil, 105, 110
Brinkley, John, 8, 39–52, 55–61, 153, 164, 174, 178. *See also* XER
Britain: acquisition of Canada (1763), 71; relations with the United States, 9, 101–02; World War I and, 120; World War II and, 5, 32, 113
British Broadcasting Corporation (BBC), 5, 94
British Information Services, 113
British Security Coordination, 113
Brooklyn, New York, 23, 28
Brundage, Fitzbaugh, 86
Bryan, William Jennings, 123, 126
Burma, 102

Cajun: communities and culture, 8–9, 65–66, 71–76, 87; French dialect of, 9, 66–68, 70–72, 75–81, 85, 87–88, 171, 178; Louise Olivier's family background and, 74–75, 82–83; sterotypes of, 66–67, 75–77, 87–89, 172, 178
Caldwell, Louis, 160
California, 26, 34, 50–51, 57; opposition to 1946 NARBA, 11, 152, 161–70. *See also* Los Angeles
Camaguey, Cuba, 107
Cambridge, Massachusetts, 113. *See also* Boston
Canada, 84; agreements and treaties with the United States, 46–47, 154–55, 159, 175–76; migration of French Canadians to Louisiana from, 8, 71–72, 171; radio interference from Mexico, 57
Canal Zone. *See* Panama Canal Zone
Capps, Robert W., 34–35, 37, 186n44
Cárdenas, Lázaro, 49, 58
Castle, William, 2, 42, 49
Chase and Sanborn Hour, 22
Chicago, Illinois, 16, 23, 51, 113, 166
Chicago Daily News, 52
Chicago Daily Tribune, 120
Chile, 4, 106, 134
China: American film exports to, 175; American radio and, 5, 92; Basic English taught in, 101–2, 112; Voice of America broadcasts to, 174
Chinese (language), 103, 113
Chiriquí, Panama, 127
Churchill, Winston, 102
Cincinnati, Ohio, 50–51
Civic Music and Art Association of Los Angeles, 18
Clark, J. Reuben, 45
Clay Center, Nebraska, 52
Clear Channel Broadcasting Service (CCBS), 160–66
clear channel stations and frequencies: Cuba's demands for, 153, 157–59; interference from foreign stations on American, 51, 161; Mexico's demands for, 47–48, 50; in the United States, 21–22, 46, 153–

54, 166–67; US-Canada agreement addressing, 46. *See also* American System of Radio Broadcasting; Clear Channel Broadcasting Service; North American Regional Broadcasting Agreement
CMQ, 168
Coahuila, Mexico, 48
Coca-Cola, 47
Coffman, G. W., 39, 41–42
Cold War, 4, 33, 173
Colombia: border with Panama, 126, 129; control of Panama by, 144, 210n33; Panama's independence from, 118, 122, 124, 128, 134
Colón, Panama, 121, 127, 144
Colorado, 50–51
Columbia Broadcasting System (CBS), 15, 21–22, 27, 32, 92
Columbia University, 92
Columbus, Georgia, 52
commercialization of American radio: *See* American system of radio broadcasting
Communications Act (1934), 21
Congress, United States. *See* United States Congress
Connecticut, 32
Conway, Burrell, 55
Costa Rica, 108
Council for the Development of French in Louisiana (CODOFIL), 88
Craig, Edwin, 160, 164
Craig County Democrat (Vinita, Oklahoma), 39
Crazy Water Crystals Company, 46
Creole: culture, 87; language, 76; people, 71, 76. *See* also Cajun
Cuba, 57; anti-Americanism and nationalism in, 153, 156–57, 160, 171; Basic English broadcasts to, 105, 107, 109; broadcasting in, 108, 131, 155, 168; education in, 109; NARBA revision demanded by, 152–61, 164, 169, 172, 178; radio laws of, 156
Culver City, California, 34
Curtis, Charles, 45

Danbury, Connecticut, 32
Daniels, Josephus, 123, 126
Davis, Jackson, 80
Dawson, Allen, 135, 140
DeConde, Alexander, 134
Del Rio, Texas, 39, 43, 52, 58
Democratic Party (United States), 29–30, 39, 42, 134, 146, 176
Denby, Edwin, 131–32
Denmark, 102
Denver, Colorado, 50–51
Department of the Interior's Office of Education (OOE). *See* United States Department of the Interior's Office of Education
Department of State, United States. *See* United States Department of State
Des Moines, Iowa, 166
de Wolf, Frances Colt, 104–5, 159–60
Dill, Clarence, 39, 42, 49
Dominican Republic, 57, 107, 134, 154
Douglas, Susan, 52, 153–54

Ebonics, 88
Eddie Cantor, 22
education: in Cajun communities, 72, 74; ideas about immigrant assimilation and, 30–31; language instruction and, 108, 113; in Latin America, 109–110; in the United States, 52–53, 68, 88
educational broadcasting, 69–71, 73, 90–91; ideas about, 18, 20, 30, 93–94, 99; language and 1–2, 8–9, 24; stations, 16, 24. *See also* Basic English; French Radio Project; World Wide Broadcasting Foundation; WRUL
E. F. Engel, 2
El Panama America (Panama), 145
El Universal (Mexico), 48
Empson, William, 112
England. *See* Britain
English (language): American national identity and, 8, 15–20, 28–38, 172; Chinese speakers of, 92; Ebonics and, 88; the French Radio Project and, 70; global prominance of, 101–2, 173–74; ideas about immigrant assimilation and, 29–31, 34, 54–55, 177; instruction of, 67–68, 171; in Louisiana, 66, 74; Mexico's radio laws regarding, 26, 48–49; the North American Free Trade Agreement and, 175–76; perceived as civilizing tool, 96–97, 101–2; as symbol of imperialism, 48, 111–12; US radio regulation and, 35, 57. *See also* Basic English
"English-centrism," 18, 177

"English-only," 19, 30
Escondido Lemon Association, 162
Esperanto, 100
eugenics, 19, 30
Europe: broadcasting and, 90, 97, 103, 106, 113; immigration to the United States from, 19; international telecommunications and, 140; national identity (early modern) in, 27; Voice of America and, 173; World War I and, 5, 120; World War II and, 8, 24, 31–34, 58–59, 61, 98–99, 106, 113, 147
Export-Import Bank, 157

Fall River, Massachusetts, 23
Farm Bureau, 152, 161–62, 165–67
Farwell, S. E., 130
Favrot, Leo M., 80–81
Federal Communications Commission (FCC), 159; complaints to, 28–30, 35, 50, 58–60; creation of, 21; educational radio and, 69–70, 73; foreign language broadcasting monitored by, 23, 35; licensing by, 22, 57, 163, 167; networks and, 26–27. *See also* Federal Radio Commission
Federal Radio Commission (FRC), complaints to, 15, 50; creation of, 2; licensing by, 22–23; networks and, 26. *See also* Federal Communications Commission
Federal Radio Education Commission (FREC), 69–70, 73
fiber optic cables, 174
"Fireside Chats," 108
Ford, H. H., 165
foreign language(s) [languages other than English]: educators of in the United States, 2, 20; hostility toward in the United States, 37, 176
foreign language broadcasting [in languages other than English]: decline of in the United States, 22–26; hostility toward in the United States, 8, 16, 25–26, 28–38, 55; interest in, 2, 17–18, 20, 23–25
Fort Hamilton, New York, 31
Fox (television network), 114
France: international radio and, 4–5; linguistic purism in, 88; loss of Canada by, 7; protests in, 175; World War I and, 92–93, 120; World War II and, 32
Franklin, Benjamin, 18–19, 30, 177
French (language): history in Louisiana of, 66–67; spoken on American radio, 17, 29, 32; standard language ideology and, 88; 2020 revision of the North American Free Trade agreement and, 175–76. *See also* Cajun; French Radio Project
French Canadians, 175; settlement in Louisiana by, 71–72, 171
French-Creole (language), 76. *See also* Cajun: language
French Radio Project, 65–68, 71–74, 76–89, 171. *See also* educational broadcasting
"Friends of Democracy," 58
Fry, Joseph, 154

Gara Leal, Leandro, 18
Garay Diaz, Narciso, 129
Gellman, Irwin, 134
General Electric, 1, 92
geography, 2–3, 27–28, 40, 52–54
Georgia, 52, 56, 166. *See also* Atlanta, Georgia
German (language), 18–19, 29, 34, 177
Germany: World War I and, 120; World War II and, 24, 32, 59, 61, 113–114, 157
Gillette, 47
Gonzalez, Antonio, 144
Goodall, Erle, 161
Good Neighbor policy, 49, 134–35, 143, 145–46, 157
Grand Coteau, Louisiana, 74
Grau San Martín, Ramón, 157, 159
Great Depression: effect on agriculture in the Far West, 162–63; effect on US-Panama relations, 141–42, 147; radio's growth during, 27, 94; scapegoating of immigrants during, 30–31, 51
Greece, 102
Green Bay Press-Gazette, 145
Griffith, Preston H., 65, 67, 72–74, 78–81, 84–85
Grove, Walter, 32
Guadalajara, Mexico, 47
Guantanamo Bay, Cuba, 156, 169
Guatemala, 109
Guy, Raymond, 158

Haiti, 57, 71, 134, 154
Harbord, James, 2–3
Harrison, Leland, 133–34
Harvard University, 73, 90
Havana, Cuba, 57, 107, 161
Hay—Bunau-Varilla Treaty (1903), 118, 122–28, 132, 135–36, 144–46
Headrick, Daniel, 181n9
Herbert, John, 85
Herdman, R. E., 164
Hindu (language), 29
Hollywood, California, 34, 51
Hooper, Stanford C., 6
Hoover, Herbert, 49, 140
Horn, C. W., 49
HP5A (Panama radio station), 144
Hughes, Charles Evans, 2, 132, 135
Hull, Cordell, 102, 142
Hull-Alfaro General Treaty (1936), 144–48
Hutchinson, Kansas, 39

identity, 6, 9; Acadian, 66–67, 76–77, 83, 85–87; addressed in the North American Free Trade Agreement, 175; Cajun, 66, 75–77, 83; Creole, 76; "doublethink" (identity strategy), 83; national and broadcasting in the United States, 11, 15–16, 27–28, 51, 60–61; national and maps in the United States, 40, 52–56; national and the English language in the United States, 8, 15, 28, 34–38, 172, 177–78; national in early modern Europe, 27; professional, 82–83; regional (American), 11, 153–54, 163–67, 170, 178; restorative identity work and, 82–83; stigmatized, 82–83
Illinois, 1, 113
"imagined community," 27. *See also* identity: national
immigrants, 34, 171; broadcasts that targeted, 16, 20, 24–25; deportation of Mexican, 51; eugenics and, 19, 30; opposition to non-English speaking, 18–19, 25–26, 29, 37, 176–77; percentage of the population of, 35–36; settlement in Louisiana by, 71; support for Franklin D. Roosevelt, 29–31. *See also* assimilation
Immigration and Naturalization Service, United States. *See* United States Immigration and Naturalization Service
immigration law: United States 19. *See also* Johnson-Reed Immigration Restriction Act
Imperial County, California, 162
India, 102, 110
Indigenous languages, 109
International Business Machines (IBM), 94
International Radio Convention (1927), 139
international radio nights, 16, 25
International Telephone and Telegraph (ITT), 140
internet, 4, 176
Iowa, 48, 166–67
Iowa State College of Agriculture and Mechanical Arts, 68
Isham, Arthur, 164–65, 168
Italian (language), 17, 32
Italy, 32
Ivey, Kay, 176

Japan, 5, 101, 119–20, 157. *See also* Sino-Japanese War
Jefferson, Thomas, 15
Jenkins, W. F., 52
Jervis, Eugene, 161
Jett, E. K., 158–61
Johnson, A. Holmes, 27
Johnson, Hiram, 146–47
Johnson, Lyndon B., 4, 174
Johnson-Reed Immigration Restriction Act (1924), 19, 25, 30, 35–36
Jurassic Park, 175

Kansas, 39, 41, 43, 45, 55–56
Kansas City, Missouri, 43
Kansas State Agricultural College, 68
Keagy, R. C., 51
Kellogg-Alfaro Treaty (1926). *See* Alfaro-Kellogg Treaty (1926)
Kennedy, John F., 4
Kern, Stephen, 4
KFI, 11, 152–55, 157–68, 178
KFKB, 41, 43, 45
Kingston, Pennsylvania, 15, 28
KMMJ, 52
KOA, 51
Kodiak, Alaska, 27

La Estrella de Panama, 145, 150

La Nacion (Panama), 145
land grant colleges and universities, 68–70
language(s): education in, 1–2, 91, 95, 108; ideas about universal, 7, 17, 100; as marker of difference and division, 3, 36, 91–93, 177–78; sign, 101. *See also* racialization of language
languages other than English. *See* foreign language(s)
La Palma, Panama, 126, 129–30, 135, 144, 148
La Prensa (Mexico), 48
Latin America and the Caribbean: broadcasting to American audiances about, 9, 17; broadcasting from the United States to, 91–92, 102, 104–11, 172, 178; education and lingustic diversity in, 109; International Telephone and Telegraph in, 140–41; radio's development in, 117, 121, 140; Rockefeller Foundation and, 96; United Fruit in, 122, 136–38 (*see also under* Panama)
Lazarsfeld, Paul, 98
League of Nations, 93
Lefevre de la Ossa, Ernesto, 125–29
Lemmon, Walter, 90–101, 103–4, 107, 113–14, 129, 171, 176
Lend-Lease, 157. *See also* World War II
Leonard, James, 165
Like Water for Chocolate, 175
Lincoln, Abraham, 15
Lincoln, Nebraska, 51
Linguaphone Records, 108
linguistic(s), 22, 71, 76, 171; academic discipline of, 66–68, 78; diversity in Latin America and, 9, 109–10; imperialism, 111–12; intolerance and, 17–19, 28–31, 37, 177; purism, 76–78, 81, 88, 172
Logan, William Seeley, 29–31, 34, 37
Long, Huey, 84
Los Angeles, California, 11, 18, 152, 154, 206n4
Los Angeles Chamber of Commerce, 161, 165
Los Angeles County, 162
Los Angeles Times, 34, 165, 206n4
Louisiana, 59, 80, 84; Cajun population of, 8, 65–66, 71–72, 82–83; French cultural and linguistic influences in, 66–68, 71, 74–78, 83, 85; migration to and settlement of, 8–9, 66, 71
Louisiana State University (LSU), 74–76, 88–89; Acadian Handicrafts Project at, 86; Department of Romance Languages at, 66, 74, 79; Department of Speech at, 66; educational radio at, 68, 70, 78; enrollment increases at, 84; Extension Division at, 65, 67, 70, 78–79, 88; Huey Long's support of, 84; Rockefeller Foundation and, 72, 79–81, 83–86. *See also* French Radio Project
Loviglio, Jason, 12
Lyons, Kansas, 55

Mahan, Alfred, 120
Maine, 146–47
Major, John, 148
Malone-Lemmon Radio Laboratories, 94
Mann, Albert, 80, 83
Mann, Thomas, 111
maps, 40, 52, 54–55
Maricopa County, Arizona, 162
Maristany, Carlos, 157–58
Marshall, John, 73, 79, 97, 106, 111, 113
Marx, Groucho, 16, 32–34, 36–38
Massachusetts, 23, 104, 133. *See also* Boston, Massachusetts
Mawhinney, Clara Krefting, 67
McCarthy, T. L., 28–29
McPherson, Alan, 134
Mentone, California, 165
Mercer, C. H., 2
Messersmith, George, 157
Metropolitan Broadcasting Corporation, 113
Mexican-American War, 45–46
Mexican Revolution, 25, 47
Mexico, 109, 133, 156, 169; anti-Americanism and nationalism in, 48–49, 111, 171–172; broadcasting in, 44, 46–49, 108 (*see also* XER); ideas about radio and cultural exchanges between the United States and, 2, 18, 173; *Like Water for Chocolate* (movie) released in, 175; Ministry of Communications and Public Works, 45, 48–49; National Assembly of, 58; North American Free Trade Agreement and, 175–76; Supreme Court of, 57; US oil in-

dustry in, 48–49, 58; World War II and, 59, 61, 157
Middle East, 175
Milborn, Eula, 56
Milford, Kansas, 41, 43, 56
Miller-Taylor Shoe Company, 52
Milliken, Hugh, 29
Milwaukee, Wisconsin, 35
Missouri, 43, 51, 56
Mississippi, 55
Modern Language Journal, 23
Monroe Doctrine, 122
Montana, 52, 60
Morrow, Jay, 131, 135
movies, 1, 100, 174–76
Munro, Dana Gardner, 133–35, 151
Muse, Benjamin, 139
music: Acadian, 85; broadcasts of in Argentina, 108; broadcasts of in Mexico, 108; broadcasts over station XER/XERA of, 50; Cajun, 75; foreign language broadcasts of, 16, 23–24; French Radio Project and, 82; ideas about cultural exchange using, 1, 17–18, 37, 90, 96; Louise Olivier's education in, 74, 81; as marker of difference, 56, 165
Myanmar. *See* Burma

Nashville, Tennessee, 166
National Advisory Council on Radio in Education (NACRE), 71
National Association for Educational Broadcasters, 69
National Association of Broadcasters, 167
National Broadcasting Company (NBC): American system of radio broadcasting and, 15, 21–22, 27; areas of the United States reached by broadcasts of, 55; broadcasting in Mexico by, 47; broadcasts to increase understanding of Latin America, 92; broadcasts to Panama by, 144; establishment of, 21; KFI as affiliate of, 154; 1946 NARBA revision and, 158; opposition to John Brinkley from, 49; shortwave broadcasts of, 27, 144
National Council for Educational Radio (NCER), 69, 71, 73
National Council on Freedom from Censorship, 35
National Education Association, 2
National Federation of Modern Language Teachers, 2
national identity. *See under* identity
nationalism, 3, 100, 178; "civic," 30; in Cuba, 10, 156; in Mexico, 48; in Panama, 10, 128, 132, 136–38, 140, 148, 151; in the United States. 19, 30–31, 58. *See also* anti-Americanism
nativism, 19, 25, 177
Navy, United States. *See under* United States
Nebraska, 41, 51–52
Newcastle, Wyoming, 51
New Deal, 32, 163
Newfoundland, 158
New Jersey, 17, 30
New Orleans, Louisiana, 74, 84
New York (state), 28, 31, 94. *See also* New York City
New York City, 32, 71; listeners in, 28–29, 32, 58; radio stations in, 23, 50–51
New York Herald Tribune, 145
New York Times: Acadian music article in, 85; Hull-Alfaro General Treaty editorial in, 145; international radio nights article in, 17; John Brinkley article in, 52; letter to the editor protesting foreign language broadcasting in, 32; World Wide Broadcasting Foundation articles in, 90–91, 95
Nogales, Mexico, 44
North American Free Trade Agreement (NAFTA), 175–76
North American Regional Broadcasting Agreement (NARBA): 1941 agreement, 10, 40, 57–59, 152–59, 172; 1946 revision, 10–11, 152–155, 161–70, 172, 178; 1950 revision, 168
North and Central American Radio Conference (1933), 50
North Carolina, 42
North Dakota, 163

O'Connor, William, 149
Ogden, C. K., 99–102, 106, 110
Ohio, 50–51
Ohio State University, 68
Oklahoma, 39, 163
Olivier, Louise, 9, 65–67, 74–89, 172, 178
Orange County, California, 162

Oregon State College, 68
Orthological Committee (Cambridge, Massachusetts), 113
Orthological Institute: in China, 102; in London, 101

Palestinian Broadcasting Service, 5
Panama, 169, 171–72, 210n33; amateur radio in, 130–31, 149; anti-Americanism and nationalism in, 10, 118, 127–28, 132, 138–40, 148–51, 171–72, 178; broadcasting and, 129–31, 144, 150; Conservative Party of, 124; economic dependence on the United States of, 138; elections in, 127; Franklin Roosevelt's influence over US radio policy in, 141–42; Great Depression impact on, 141–42, 147; independence from Colombia of, 118, 122–124; market for US radio receivers in, 131; Ministry of Foreign Affairs of, 127, 137–38; National Assembly of, 127–28, 135–36, 139, 145; National Liberal Party of, 123–24, 127–29; 1914 presidential decree, 125–27, 131–32, 135–36, 139, 143; radio policies of, 121–25, 131–32, 136–40, 149, 151; United Fruit in, 121, 130, 136–38, 140; US military interventions in, 127; US multinational corporations in, 121–22, 136–38, 140–41; US radio policy toward, 10, 117–27, 130–51; World War II and, 147, 157. *See also* Alfaro-Kellogg Treaty; Hay Buneau Varilla Treaty; Hull-Alfaro General Treaty, Panama Canal; Panama Canal Zone
Panama Canal: Belisario Porras Barhona's views of, 124; building of, 10, 119–20, 122, 134; compensation to Colombia for, 134; Great Depression effect on traffic through, 141; opening of, 120, 125; as symbol of progress and civilization, 120, 141; US Navy radio communications system and, 121–23, 131, 137, 151; US policies for the defense of, 142–43, 146–47, 150; US relinquishment of control over, 151. *See also* Alfaro-Kellog Treaty; Hay Buneau-Varilla Treaty, Hull-Alfaro General Treaty; Panama; Panama Canal Zone
Panama Canal and Neutrality Treaty (1977), 151
Panama Canal Zone: racism of US authorities in, 135; US authority in, 10, 117–19, 121, 123; US authority claimed beyond, 123–26, 137–38; US opposition to amateur broadcasting in, 130–31, 149. *See also* Alfaro-Kellog Treaty; Hay Buneau-Varilla Treaty; Hull-Alfaro General Treaty; Panama; Panama Canal
Panama City, Panama, 117, 127, 129, 144
Panama Radio Club, 130, 149
Panama Railroad Company, 144
Pan American Union, 104
Passaic, New Jersey, 17
Pennsylvania, 15, 28, 133; Benjamin Franklin's concerns about Germans in, 18, 30, 177
Phenix Radio Company, 54
photography, 100
Piedras Negras, Mexico, 57
Poland, 24, 59
Polish immigrants, 24
Porras Barahona, Belisario, 123–28
Portuguese (language), 17, 106, 110
Potter, Simon, 2
Prescott, Richard D., 149
Price, William Jennings, 127, 129
Princeton Radio Research Project. *See* Radio Research Project
Princeton University, 93
Progressive Party (United States), 146
Puerto Obaldía, Panama, 126, 129–30, 135, 144, 148
Puerto Rico, 29
Punto Mala, Panama, 126

Quebec, 175

race: accents and perceptions of, 3, 6, 109; discourses of, 3, 6, 18, 25, 125. *See also* racism
racialization of language, 18–19, 25, 83, 177–78
racism: accusations of, 177; in American society, 17–19, 25, 31, 83; in *Amos 'n' Andy* radio program, 6, 22; eugenics and, 19, 30; of US official in Mexico, 47; of US official in Panama, 135; the US-Mexican border and, 165
Radio Act (1927), 21
Radio and the Printed Page, 98

Radio Broadcast, 53
Radio Communications Convention between the United States and Panama (1936), 144–46, 148. *See also* Hull-Alfaro General Treaty
Radio Corporation of America (RCA), 2
Radio-Craft, 24
Radio Daily, 159
Radio Habana, 155
Radio Industries Corporation, 94
Radio News, 20, 54, 55
radio receiver ownership: in Cajun communities, 72, 81; in Europe, 97; in Latin America, 97, 105; in Mexico, 47; in Panama, 117, 131, 148; in the United States, 20–22, 47
Radio Research Project, 73, 98
Redfield, South Dakota, 60
Redlands, California, 164–65, 168
Regional Broadcasters Committee, 167
regionalism. *See under* identity: regional
Reilly, Henry J., 146
Republican Party (United States), 2, 134, 146, 176–77
restorative identity work. *See under* identity
Rhode Island, 32
Rice, Martin, 1
Richards, I. A., 98–104, 111–12
Rio Grande, 39, 48, 51–52, 59
Risch, O. C., 17
Riverside County, California, 162
Rockefeller, John D., 69
Rockefeller Foundation, 8; Basic English and 100–102, 112; the French Radio Project and, 65, 69–73, 78–86, 88; the World Wide Broadcasting Foundation and, 95–99, 105–7, 111–13
Roosevelt, Franklin, D., 58; Basic English supported by, 102; Good Neighbor policies and, 49–50, 101, 134, 157; Fireside Chats of, 108; Japanese-American internment and, 31; New Deal policies and, 32; 1932 election of, 141; Panama radio dispute and, 141–43, 147–48; political support from immigrant groups for, 29–30
Roosevelt, Theodore, 15
Russia, 120. *See also* Soviet Union
Russo, Alexander, 55
Ryan, William, 161
San Antonio, Texas, 166
San Bernardino County, California, 162
San Diego County, California, 162
San Francisco, California, 206n4
San Francisco Chronicle, 206n4
San Jose, Costa Rica, 108
San Luis Potosí, Mexico, 47
satellites, 4, 174
Schreiber, Frank, 168
Scientific Monthly, 94
Seagrave, Orville, 24
Short-Wave Broadcasting Corporation, 90
shortwave broadcasting: educational programs, 9, 90, 98; NBC use of, 27, 92, 144; satellites and fiber optic cables compared to, 174; technological development of, 94–95; as US foreign policy tool, 97, 150. *See also* World Wide Broadcasting Foundation
shortwave receiver ownership: in Europe, 97; in Latin America, 97, 107, 110
Singapore, 110
single dial tuning, 94, 176
Sino-Japanese War (1937–1945), 101–2, 147. *See also* World War II
650-mile rule. *See* clear channel stations and frequencies
Smith, Harley, 9, 65–68, 70–73, 76–85, 87–89
Smith, James Monroe, 78–80, 84
social construction of technology, 7, 98
social media, 176
social sciences, 30
South, John Glover, 129–30, 135
South Dakota, 51, 60, 163
Southern Broadcasting Network (Louisiana), 70–71, 78
Soviet Union, 4; 1939 invasion of Poland, 24. *See also* Russia
Spain: Cuba as colony of, 153, 155; Louisiana as territory of, 71; protests in, 175
Spanish (language), 133; broadcasts in, 11, 17, 26, 55–57, 104; in Latin America, 104–6, 109–10, 171–73; in Louisiana, 71; as marker of difference, 29, 38; racialization of, 18, 25–26, 83, 177–78
Spanish (people and culture): Creole identity and, 76; influence in American culture, 96
Spanish-American War. *See* War of 1898

Spearman, Paul, 167
Staff, Hector, 150
standard language ideology, 77, 88
Standard Oil, 69
Star and Herald (Panama City), 117, 129, 144
State Department, United States. *See* United States Department of State
St. Clair, K., 31, 34
Steetle, Ralph, 80–81, 84
Stewart, H. C. R., 56
Stewart, H. S., 152
Stimson, Henry, 140–41
Stirling Jr., Yates, 146
St. Louis, Missouri, 51
Sullivan, Sylvester, 28–29
Swarthmore College, 113
Sweet, George, 2, 173
Sydney, Montana, 52, 60

Tallapoosa, Georgia, 56
T'an, Pin Pin, 102–9, 113
Taylor, T. E., 52
technological determinism, 6–7, 98
telegraph, 3–4, 118, 122
television, 33, 114, 175
Tennessee, 166
Texas: agricultural production in, 162; "Border Strike Force" in, 177; Crazy Water Crystals based in, 46; Farm Bureau in, 166; John Brinkley relocation to, 39, 43, 52, 58; Spanish language broadcasting and, 26, 57; as territory of Mexico, 45
Time, 52
Tonosí, Panama, 210n33
Topeka, Kansas, 56
tourism, 26, 156
Trans-Isthmian Highway, 144
Trapp, Charles, 56
Treaty of Versailles (1919), 93. *See also* Versailles Peace Conference
Tropical Radio and Telegraph Company, 136–37, 140–41. *See also* United Fruit
Tufts University, 90
Tyler, Charlotte, 104, 113
typewriter, radio operated, 94

United Date Growers Association, 161–62, 164
United Electric Railway Company, 56
United Fruit, 121–22, 130, 136–38, 140
United States: acquisition of Louisiana Territory, 71; acquisition of territory from Mexico, 45–46; border with Mexico, 26, 40, 44–61, 165, 172, 176–77; Civil War, 3; Cold War and, 4, 173; education in, 52–53, 68, 88; radio agreement with Canada (1932), 46 (*see also* North American Regional Broadcasting Agreement); North American Free Trade Agreement and, 175–76; radio receiver ownership in, 20–22, 47; tourists in Cuba from, 156
United States Army, 123, 145–46, 149; National Guard, 176
United States-Canada Free Trade Agreement (1987), 175. *See also* North American Free Trade Agreement
United States Congress, 19, 21, 102, 122 168. *See also* United States Senate
United States Department of State: border radio dispute with Mexico and, 39, 42, 45, 49–50, 60; radio dispute with Cuba and, 152, 159–61, 164–65; radio dispute with Panama and, 119, 123, 126, 132–33, 135–36, 138, 140, 142–49, 151; support for World Wide Broadcasting Foundation from, 104–5, 113
United States Department of the Interior's Office of Education (OOE), 69–70, 73
United States Immigration and Naturalization Service, 113
United States Internal Revenue Service, 59
United States Navy (including Department of), 2, 6, 90, 92; base at Guantanamo Bay, 156, 169; Panama radio and, 118–19, 122–23, 126, 128–29, 131–33, 136–51
United States Postal Service, 59
United States Senate, 49, 57, 93, 145–48, 160, 164, 168
University Broadcasting Council, 71, 73
University of Chicago, 113
University of Illinois, 1
University of Kansas, 2
University of Missouri, 56
University of Wisconsin, 1
USS George Washington, 92, 94, 129

Vaillant, Derek, 4, 23
Valdes, Hector, 128

Vallance, William R., 140
Valparaiso, Chile, 134
veillées, 75. *See also* Cajun: culture
Ventura County, California, 162
Vera Cruz, Mexico, 109
Versailles Peace Conference (1919), 92–93
Vicksburg, Mississippi, 55
Villa Acuña, Mexico, 39, 43, 48
Vinita, Oklahoma, 39
Voice of America, 113, 173–74
Voice of Panama, 144
Voltaire, 111

War of 1898, 156
Washington (state), 39
Washington, George, 15
WCFL, 16
Welles, Sumner, 142–43
WGN, 50–51, 168
White, Francis, 135, 141
White, Wallace, 146, 148
WHO, 166
Wilbert, Ida Lee King, 68
Williams, Alonzo, 56
Williams Jr., James T., 146
Wilson, Woodrow, 15, 92–94, 114, 129, 143
Wisconsin, 1, 35
WLS, 166
WLW, 50–51, 166
Woods, Bryce, 134
W1XAL, 90, 95. *See also* World Wide Broadcasting Foundation; WRUL
WOR, 50–51
World War I, 1, 3, 35, 53; alliances in, 120, 126; broadcasting's origins and, 129–30; eugenics and, 19; as motivation to create Basic English, 99–100; as motivation to establish the World Wide Broadcasting Foundation, 92–94, 99, 114; outbreak of, 4, 121; US Navy authority over radio and, 150; Walter Lemmon's service during, 90, 92
World War II, 8, 58, 61, 106, 119, 149, 171–73, 181n9; American concerns about foreign language broadcasting during, 31–36; international radio during, 5; internment of Japanese-Americans during, 31; Lend Lease program and, 157; outbreak of, 24, 59, 98–99, 114, 147; radio broadcasting and regionalism in the United States during, 154; United States entry into, 25, 36, 59, 113, 157, 163–64, 172
World Wide Broadcasting Foundation (WWBF), 90–91, 94–102, 104–7, 110–14, 173. *See also* Walter Lemmon; WRUL
WRUL, 95, 103–105, 113–114. *See also* W1XAL; World Wide Broadcasting Foundation
WSAR, 23–25
WSB, 50–52, 56
WSM, 160, 166
Wyoming, 51

xenophobia, 16–18, 32
XER/XERA, 39–41, 43–45, 48–53, 55–57, 59–61

Yenching University, 102
Yiddish, 29
Yonkers, New York, 28

Zionism, 5

MICHAEL A. KRYSKO is an associate professor of history at Kansas State University. He is the author of *American Radio in China: International Encounters with Technology and Communications, 1919–41*.

The University of Illinois Press
is a founding member of the
Association of University Presses.

Composed in 10.25/13 Marat Pro
with Trade Gothic LT Std display
by Lisa Connery
at the University of Illinois Press

University of Illinois Press
1325 South Oak Street
Champaign, IL 61820-6903
www.press.uillinois.edu